Psychological Operations and Political Warfare in Long-term Strategic Planning

Psychological Operations and Political Warfare in Long-term Strategic Planning

Edited by
Janos Radvanyi

PRAEGER

New York
Westport, Connecticut
London

Library of Congress Cataloging-in-Publication Data

Psychological operations and political warfare in long-term strategic planning / edited by Janos Radvanyi.

p. cm.

Includes bibliographical references.

ISBN 0-275-93623-6 (alk. paper)

1. Psychological warfare. 2. Strategy. 3. Soviet Union—Military policy. 4. United States—Military policy. I. Radvanyi, Janos.

UB275.P77 1990

327.1′4—dc20 90-31183

Library of Congress Catalog Card Number: 90-31183
ISBN: 0-275-93623-6

First published in 1990

Praeger Publishers, One Madison Avenue, New York, NY 10010
An imprint of Greenwood Publishing Group, Inc.

Printed in the United States of America

∞

The paper used in this book complies with the Permanent Paper Standard issued by the National Information Standards Organization (Z39.48-1984).

10 9 8 7 6 5 4 3 2 1

Contents

Acknowledgments

The research unit of the Center for International Security and Strategic Studies at Mississippi State University is pleased to present its latest anthology *Psychological Operations and Political Warfare in Long-term Strategic Planning*. This volume is the result of an extensive research and discussion effort from the best minds of the fields of psychological operations and political warfare. Our contributors from Canada, Great Britain, France, Germany, and the United States enriched the volume.

As we reiterate our indebtedness to all our contributors for their generous assistance and stimulating thoughts, we owe a special thanks also to Congressman G.V. "Sonny" Montgomery for his advice and encouragement.

In preparing the manuscript, we were crucially dependent upon the unfailing assistance of Jean H. Johnson, Center for International Security and Strategic Studies. In this respect, we are further indebted to Tan Tsai, Julie S. Stevenson, and Samantha Steeber.

Finally, we wish to express our thanks to the Bradley Foundation, which provided financial support for the research and the preparation of the book. Our gratitude also goes to the staff of Praeger Publishers, especially to Editor Dan Eades for seeing the book through publication.

Psychological Operations and Political Warfare in Long-term Strategic Planning

1

Introduction to Psyops

Janos Radvanyi

Psychological operations are as old as the history of mankind; indeed, the Bible records numerous cases, such as the "torch-game" deception of Gideon against the Midianites that forced them to retreat. The Athenians' clever use of psychological warfare against the Melians in the Peloponnesian War is well documented, as is the Romans' disinformation and blackmail campaigns that ultimately forced the Carthaginians to surrender in the Punic Wars. And, one should not forget that the success of the Golden Horde of Genghis Khan partly depended on frightening rumors about the savagery and the greatly exaggerated numbers of the Khan's horsemen army. Manipulation of the adversary through psychological operation and special warfare continued throughout the Middle Ages, the Renaissance, and the Age of Absolutism. Ready-made recipes for conducting psychological warfare and special operations (psyops) can be found in the writings of Niccolo Machiavelli and his contemporaries.

In modern times only psyops techniques have changed; their goal of defeating an opposing power remains. No better example can be found than Otto von Bismarck's "editing" of his ambassador's famous Ems telegram, a falsification which ultimately provoked the French into the 1870-71 war they were almost certain to lose. Totalitarian regimes, such as Fascist Italy, Nazi Germany, and militaristic Japan, widely used psyops in the Abyssinian War, in the Spanish Civil War, against Czechoslovakia and Poland, and in Manchuria and the rest of East Asia. Among the many well-known cases, the Third Reich's forgeries, which led to the 1932 execution of the best military strategist of the Red Army, Marshall M. Tukchachevsky, are an often cited example.

Throughout its history, the U.S. approach toward practicing psyops can only be characterized as mild, if not timid. Distributing leaflets to invite British troops to desert during the War for Independence was perhaps the first hesitant attempt. For a long time thereafter, the politicians in Washington considered psyops too dirty and too ineffective to practice. Not until World War II did the newly organized Office of Strategic Services (OSS) carry out covert black propaganda, together with the Allies, on strategic and tactical levels. During the Cold War, the U.S. government was more interested in containing Soviet advances and providing technology to deter nuclear war than in using psyops to roll back the Iron Curtain—or at least punch holes in it. The dissemination of black propaganda was almost completely halted in the mid-1970s and 1980s when numerous congressional investigations curtailed CIA covert operations. Only special techniques associated with the preparation for low intensity conflicts were developed through the issuance of the Defense Guidance of 1981 which directed all the armed services to develop a Special Operation Forces (SOF).

Since its establishment, no power has used psychological operations and special warfare more extensively nor more successfully than the Soviet Union. Perhaps, because of their Mongol heritage, the Bolsheviks have been masters in the arts of deception, disinformation, blackmail, and black propaganda. As a matter of fact, Russian Marxist revolutionaries' interest in manipulating their adversaries through disinformation and treachery can be traced back to 1903, when the young Vladimir I. Lenin captured the party leadership at the Second Russian Social Democratic Workers' Party Congress while claiming the label "Bolshevik" (from the Russian *Bolshinstvo* or majority) when, in reality, he was in the minority. His fraudulent opportunism provided a wide base on which to build an elaborate system of deception, espionage, and blackmail for his future regime.

From the October Revolution of 1917 to the consolidation of power, the Communists (as the Bolsheviks renamed themselves in 1918) widely used psyops against the Whites and their temporary allies, the leftist SRs. Lenin's replacement, Stalin, did not hesitate to direct his secret police, the GPU, and later the NKVD, as well as the Communist International and its successor the Communist Information Bureau (the Comminform), to engage in protracted deception campaigns against the capitalist world. Stalin's molding of Eastern Europe into a satellite bloc was partly accomplished by the use of large-scale psychological warfare. But his special operations—sabotage, blackmail, and political pressure—against the "heretic" national communist Yugoslav Broz Tito were perhaps the most spectacular.

With the communist victory in China, the Korean War, and North Vietnam's invasion of the South came a new form of psyops—"brainwashing." The Chinese call it "thought reconstruction;" the West named it "pseudo-psychiatric twisting of personality." No matter what the label, it was used quite effectively to force POWs not only into submission but also into working as propagandists against their own countries. When the first "reformist" Soviet party leader, Nikita Khrushchev, called upon the "progressive forces" of the world to unite in their struggle for "peaceful coexistence," a qualitative change in Soviet psychological operations came about. Parallel with this "peace offensive," Khrushchev's newly established (in 1959) special unit for black propaganda and disinformation within the KGB, the so-called Department for Active Measures, carried out anti-American campaigns. The building of the Berlin Wall, the risky and expensive blackmail in nuclear diplomacy known as the Cuban missile venture, and the propaganda war around the Sino-Soviet rift were milestones of Khrushchev's special operations.

The period of stagnation under Leonid Brezhnev, as it is characterized today by his successors, was perhaps the most dynamic as far as psyops were concerned. With the assistance of Soviet advisers, the Departments for Active Measures in all European Satellite countries and Cuba conducted intensified and coordinated psyops campaigns. Throughout the world, their black propaganda aimed to discredit the United States and its Allies and manipulate the Western press and electronic media. The 1968 Warsaw Pact countries' invasion of Czechoslovakia and the subsequently proclaimed Brezhnev Doctrine "to save socialism" were used as new means to quell any anti-Soviet, anti-communist uprisings in Eastern Europe. And lastly, while the United States was stricken by Watergate and the Vietnam syndrome, the Soviet Union was busy extending well camouflaged, low intensity operations in Asia, Africa, and Central America. Using Cuban, East German, Czech, North Korean, and Vietnamese surrogates, and Soviet military hardware, the Brezhnev leadership put pro-Moscow communist groups into power in Laos, Cambodia, Angola, Ethiopia, Nicaragua, and Afghanistan. Moreover, substantial Soviet support, coupled with active measures, promoted successful Libyan- and Syrian-led insurgencies, guerilla warfare, and wide-range terrorist activities.

In retrospect, however, the colonialist expansion and influence peddling of Brezhnev in the 1970s resulted to some extent only in greater insecurity and less strategic advantages for the Soviet Union. The huge expenditure necessitated by its global military buildup further weakened the economy of the already crumbling Russian empire. Threatening Western Europe with medium range SS-20s triggered Western countermeasures and clearly impacted unfavorably upon the Soviet military posture. Direct and indirect Russian involvement in regional conflicts, like the wars in Afghanistan, Ethiopia, Cambodia, and Angola, were not worth the heavy investment and the risks involved. On the home front, the failure of the communist system became even more apparent as the neo-Stalinist Brezhnev regime widened the rift between the party and the society.

Thus, and not surprisingly, Mikhail Gorbachev's rebuilding, *perestroika*, of the political and economic system, revitalization of the society by openness, *glasnost*, and chartering of *new thinking* in foreign policy were proposed to assure the survival of the Soviet Union. But, the widely applauded reforms of Gorbachev amounted to, in reality, a

Soviet retreat equal in dimensions to Lenin's signing of the humiliating Brest-Litovsk Peace Treaty with the German Empire in 1918. In each case these great gambles were not taken to liquidate the system. Lenin's gamble gave him *peredyshka*, a breathing space to recover; in contrast, Gorbachev's gamble could fail and the whole system collapse.

There are strong evidences that such a scenario could happen. So far, the ethnic rebellions in the Central Asian Republics of Azerbaijan and the Ukraine and the recurring widespread labor unrest are controlled by Gorbachev's allies, the internal security forces and the military. But, succession movements in the Baltic Republics of Estonia, Latvia, and Lithuania are only contained with great difficulties and the outcome is uncertain. Meanwhile, his overhaul of the party apparatus and the ouster of anti-perestroika elements from the government bureaucracy has been accomplished. And, Gorbachev has introduced new presidential powers to keep the momentum going. Thus, in these instances the gamble is working.

On the other hand, his other big gamble, Eastern Europe, has not worked. The East European socialist regimes ceased to exist. Poland, East Germany, and Hungary have already elected parliaments where even the reformist communists have lost and the governments have no communist members. Czechoslovakia is moving at an accelerated pace toward democracy. The Warsaw Pact is in shambles. Czechoslovakia and Hungary have reached agreements with the Soviet government to remove all Red army units from their countries in 1991. East Germany, Poland, Czechoslovakia, and Hungary are on the verge of quitting the Soviet-dominated Trade Council for Mutual Economic Assistance (Comecon). The revolution triggered tremendous explosions in Bucharest, and the Bulgarian communists were also forced to introduce reforms. The domino effect of the unanticipated revolution of 1989 reached dogmatic Albania, communist South Yemen, North Korea, and Cuba and even one of closest allies of the USSR, the People's Republic of Mongolia. For all practical purposes, the communist models, including the Gorbachavian reforms, are no longer examples to be imitated.

Under such circumstances, psychological operations and political warfare become even more important for Gorbachev than at any other time in the course of Soviet history. Obviously, the forms and methods of today's active measures are quite different from those of the Stalinist or Brezhnevist periods, as Gorbachev's psyops demonstrated in Lithuania and Estonia. Abroad, the use of harmless, nonpolitical front organizations' propaganda campaigns coordinated through foreign nonruling leftist parties have become more sophisticated. The reported Soviet $4 billion annual budget slated for disinformation is used to stir up anti-Western, especially anti-American, sentiments around the world as before. The advent of glasnost did not curtail Soviet intelligence activities either. According to CIA Director William Webster, Moscow has enjoyed more success with its intelligence operations in the last four years than at any time in postwar history. What is new is that Soviet open and covert propaganda successfully projects a less threatening image of itself in order to slow down Western defense expenditure and gain the needed Western credit and technology. It is also new that the industrial espionage effort of the KGB has been engaged in intercepting unencrypted telefaxed communications, blueprints, and documents of defense-related companies worldwide. On the political front Gorbachev's innovation centers around the creation of a "Common European Home" (with Soviet participation) and the dissolution of the North Atlantic Treaty Organization (NATO), or at least its denuclearization.

Against this background, the present anthology offers a detailed analysis of the rapidly changing Soviet strategy of psychological warfare and special operations and contains ample suggestions on how the United States and its allies could or should adjust to these new circumstances. In this respect, one of the best experts in Soviet military affairs, Army Lt. General William Odom (ret.), argues that in reality the terms "political warfare" and "long-term strategic planning" have little use or meaning in the United States. The first seems to contradict U.S. democratic principles of fairness and ethics; the other is impractical and almost impossible, given the U.S. political system. While there have been brief spans of a coherent U.S. national strategy, usually a general lapse back into a reactive stance followed. The Soviets, on the other hand, have been successful in implementing long-term strategy and psyops. General Odom sees the nature of current Soviet politics not as a diminished or erased threat, but rather as a redirection of strategy as a result of a realistic appraisal of the Soviet Union's poor economic performance and changes in external factors. He concludes that the West's response to the new Soviet strategy should be to continue the qualitative military and regional competition while advancing policies tied to real and measurable reforms in Soviet policies.

Dr. Alfred Paddock, Jr., U.S. Army Colonel (ret.), one of the most renowned authorities on psyops, provides a history of U.S. military psychological operations with its peaks and valleys. While he is critical that psyops has been relegated to an orphan-like status in the peacetime armed forces, he pays tribute to the Reagan administration's decision to revitalize the U.S. military psyops capabilities in terms of doctrine, planning, education, and force structure modernization. In conclusion, however, Dr. Paddock cautions that the perception of a diminished Soviet threat could cut short the improvements.

Mr. Herbert Romerstein from the United States Information Agency gives an exciting account of the inner workings of Soviet active measures. He classifies Soviet active measures into covert, overt, and semi-covert areas, all designed to influence and destabilize the international policies of the United States and its allies. While these activities date back to the early days of the Bolshevik Revolution and include the personal involvement of Lenin, significant redirection and reorganization occurred in the Gorbachev era. The focus is now to capitalize on Gorbachev's public relations success and concentrate on strengthening the newly improved Soviet image in the international community. Mr. Romerstein offers an interesting and valuable chart detailing the reorganized and highly structured Soviet Department for Active Measures and lists the names and positions of its commissions.

As a logical follow-up of Mr. Romerstein's study, the director of Canada's Mackenzie Institute, Dr. Maurice Tugwell, a theoretician of psyops, examines terrorism from both historical and theoretical perspectives. He views terrorism as a subsidiary form of revolutionary warfare and stresses the ultimate purpose of terrorism as being persuasion or the changing of people's minds. In addition, he sees democracies, because of their very nature, as particularly vulnerable to terrorist acts. He forecasts that the Soviet Union, under Gorbachev, will publicly denounce terrorism while continuing its covert support, though to a lesser degree than in the past.

Dr. Josef Joffe, the internationally known and respected foreign editor of one of West Germany's largest and most influential newspapers, *Suddeutsche Zeitung*, offers an important analysis of the Soviet psychological warfare campaign against West German public opinion. The Soviets have taken a "carrot and stick" approach, he writes, fluctuating between overt threats and promised rewards. His basic thrust is that, for whatever reasons, West German public opinion has been dramatically transformed so as to be highly favorable to Soviet goals. He stresses, however, that the basics of the population's allegiance to NATO and the presence of U.S. military forces has not changed. He attributes the rejection of nuclear weapons to the changing German perception of the Soviet threat.

Professor Teresa Rakowska-Harmstone provides a case study of the Soviets' imposition of communist rule in Poland by focusing on the use of psychological warfare against the Polish population. While pointing out both successes and failures, she asserts that in Poland the overall result is failure—a conclusion validated by the events taking place in the 1980s. Characterizing the post-World War II communist government in Poland as a "parachuted" communist regime, she outlines the process by which they attempted to undermine and totally destroy the fabric of independent Polish society and politics. Analyzing the major themes of historical perspectives and attitudes toward the social system, she concludes that, on the whole, the results have been counterproductive to Soviet aims. The situation in Eastern Europe today adds credence to her conclusion.

Admiral Pierre Lacoste (ret.), of France, the former head of the French Foreign Intelligence Service, presents an historical perspective on military involvement in psychological warfare based on the French experience of psyops during World War II, the Algerian War, and in Indochina. He emphasizes the long history of psychological warfare and the difficulties it creates for the individual combatant. He reminds us that it is fear, not virtue, that prevents nations from going to war. He therefore warns against drastic reductions in military strength in response to perceptions of a diminished threat and urges policies based on realism rather than idealism.

The picture of modern day psyops would not be complete without the in-depth study of psychological warfare in Third World insurgencies offered by Dr. Michael Radu, a leading researcher at Philadelphia's Foreign Policy Research Institute. He discusses the techniques and methods of "propaganda by deed," terror, and nonviolent propaganda of Third World insurgency politico-psychological operations against the United States. Dr. Radu describes how the KGB Active Measures Department uses political and mass front organizations and solidarity groups in their hostile disinformation campaigns and underlines that due to the sophistication of contemporary politico-psychological warfare, most Western or pro-Western governments are ill-prepared to cope with its demands.

All in all, both the editor and the contributing authors were determined to offer an objective and realistic view of modern day psyops and were especially concerned not to hammer out consensus positions in this anthology. They have identified, analyzed, and evaluated facts for the unique purpose of understanding and clarifying issues and problems affecting vital security interests of the United States and its allies. Though the individual chapters do not claim to furnish all the answers, it is hoped that they will at least provide new dimensions for the consideration of Soviet psychological operations in the Gorbachev era.

2

Psychological Operations and Political Warfare in Long-term U.S. Strategic Planning

William E. Odom

INTRODUCTION

In principle, I do not like the topic of *psychological operations* and *political warfare* (psyops) in *long-term strategic planning*. If it is what it appears to be, concerned primarily with psychology, then I am wholly unqualified to write about it. If it is political warfare, that seems redundant to a student of Clausewitz, and it seems unethical to a student of democracy. If it is primarily about long-term U.S. strategic planning, a student of the record would probably argue that there is no such thing. Americans are intellectually ill-disposed toward it, and our system makes it impossible in any case. We are given to ad hoc responses to strategic threats, and our system of policy-making and resource allocation for foreign and military policy gives the advantage to a pluralism of competing bureaucratic, economic, and political factions, making long-term planning a fanciful aspiration, not a serious undertaking.

Following are some wide ranging observations, some critical but preliminary thoughts, intended to explore the topic, to ask if there is something useful that we can discover. They may cause you to think more critically about the issues involved.

PROBLEMS, CATEGORIES, AND CONFUSION

What are the origins of the terms psychological operations and political warfare, and what do they mean to most government officials? The two places in the U.S. government where they have an institutional base are the U.S. Information Agency (USIA) and the U.S. Army. USIA is well known. It works under many restraints, and it is seldom included in National Security Council planning and policy-making. It does its best to create a favorable image of U.S. policy through the spread of information. In most circles it is not taken very seriously because the commercial media have a far greater role in informing the world about U.S. policy. It is hardly a match for NBC, CBS, and ABC, not to mention many publications and news agencies.

In the U.S. Army there are psyops units with loudspeakers and printing capabilities. These organizations are essentially the residual structure from such operations in World War II where the target was enemy soldiers under immediate U.S. attack. Psyops also enjoyed a certain revival during the Vietnam War when we wanted to win the hearts and minds of the Vietnamese citizenry. Today they receive some support for application in so-called low-intensity conflicts. Critics of these Army efforts have a solid point when they argue that the Army is ill-prepared to do the proper political analysis that everyone agrees is necessary as a basis for developing effective psyops tactics. Moreover, even if the Army did an effective political analysis, it does not have control of all the U.S. agencies involved in a place like El Salvador. Many of these agencies pursue policies that are not based on a well conceived U.S. strategy, and they are not about to take orders from the Army to get in line with Army psyops.

A couple of other efforts are worth mentioning. The new National Endowment for Democracy is a more sophisticated attempt to revive psyops at the strategic level. And, of course, Radio Liberty and Radio Free Europe have long been in operation. RFE may be the most effective psyops/political warfare effort developed by the United States since World War II.

This, then, is the present state of institutionalization of psyops and political warfare. What do the terms mean to most officials? For most, they conjure up the notion of trying to swindle the opposition, to confuse or mislead them. For many, there is something un-American, something unfair about such efforts. Others do not oppose them in principle; they just doubt that we are very good at them. Thus, they do not support them.

Long-term strategic planning is a little better off in our government, but not a lot. More people are for it. It does not suffer the odium that surrounds psyops and political warfare. Rather, no single institution in the government seems to have adequate control over policy to implement it. Samuel Huntington told a conference at the National Defense University in 1988 that we have some prospect of developing a military strategy but that a national strategy, broader than military, seems beyond our system's capability. I worked with him in trying to develop such a strategy during the Carter administration with PRM-10. The policy document that resulted, PD-18, was 95 percent military. We could not even get the executive branch to stand solidly behind a national strategy, not to mention bringing along the Congress.

If there is to be any serious improvement over the present state of affairs, I am inclined to believe that we should start by abandoning the terms psyops and political warfare in favor of a return to the basic categories of U.S. politics and warfare. What is our purpose in the world as a nation? What do we want? What are we about? If we start here, we may make modest progress.

AMERICAN PURPOSE AND THE WORLD

Two deep and opposite feelings about the rest of the world can be traced in American history. On the one hand, we believe we are special, that our democratic system is unique, that the rest of the world is just unfortunate and cannot become like us. That view guided George Washington in his farewell address, when he warned us to avoid foreign entangling alliances, and prevailed throughout the nineteenth century.

In the Spanish-American War, however, we asserted our power beyond our shore. Implicitly we abandoned Washington's view. It still had a strong constituency, however, persuading President Wilson to run on a promise to keep us out of the war in Europe. But there was, on the other hand, a feeling among some Americans that we should spread our political blessings to the rest of the world. Wilson shared it, and we soon went to war to end all wars. What started with Teddy Roosevelt and "big stick diplomacy" was converted by Wilson into a new internationalism for peace and self-determination, expressed in his Fourteen Points.

You know the record thereafter. We rejected Wilsonian internationalism for a couple of decades but then reasserted it vigorously, not only in World War II but also in the postwar international order, and we have given the upper hand to this strain of U.S. political sentiment ever since.

Why have we reversed our traditional stance vis-à-vis the world so fundamentally? Have we rejected our initial political moorings and values? What has been our purpose in this change? Marxists, dependency theorists, and critics try to characterize us as the new bastion of imperialism and colonialism. They cite business interests and cultural imperial tendencies.

The true answer is much simpler. We entered both world wars because we genuinely believed that democracy abroad was in danger, and we were concerned that if it perished there, it would be more difficult to defend in America. Our commercial expansion did not come first. It has come after, in the wake of defending democracy abroad. Latin America is an exception to this sequence, but even there the democratic impulse was not wholly subordinate to commercial concerns. The freest elections ever held in Nicaragua were in 1926 under U.S. military intervention.

The historical record indisputably justifies the American judgment. Democracy has been in danger. The major threat has been from totalitarianism. We have seen it in two variants, the radical right,

Nazism and Fascism, and the radical left, Leninism and communism.

This historical record also justifies the efficacy of U.S. abandonment of isolationism in favor of assertive defense of democracy abroad. West Europe and East Asia have enjoyed their longest period of peace in modern history under U.S. leadership.

We may argue about how we could have been more effective, but it should be beyond dispute that our purpose was justified, and the results have been positive. If we have done well, then, what is the concern?

Since the Vietnam War, voices of doubt about the rightness of our policy have been strong. Many of our intellectuals and media elites, even some of our politicians, seem to believe that we have created what we set out to prevent, new kinds of militarism and imperialism. The Soviet invasion of Afghanistan and the record of the Soviet military buildup, however, gave this line of argument a serious setback. It has been on the defensive throughout the Reagan years. Why are we again concerned with it?

The answer is fairly obvious. Gorbachev's domestic policies coupled with his new peace offensive toward the West have convinced many people that the Cold War is over, that communism is finished, that the original purpose of our new globalism is no longer justified. The threat to democracy abroad is in irreversible decline.

THE CHANGING NATURE OF THE THREAT TO LIBERAL DEMOCRACY

For those who are skeptical about the declining threat and who do not believe our purpose is accomplished, psyops and political warfare easily come to mind as instruments for continuing the struggle. Such a reaction is understandable, but it is not likely to be effective if it falls back on these instruments of competition as they are understood in the American context.

They are too easy for critics to ridicule; they seem politically unsophisticated. As they have developed institutionally in our system of government and strategic planning, they are rather feckless. Domestic critics can have a field day with them. Even if we use them more vigorously, they will not enlighten the peripatetic Europeans who flock to our shores to tell us that Leninism is dead, that the Cold War is over, that U.S. policy is the real menace to peaceful evolution toward democracy in East Europe. These Europeans and the numerous American enthusiasts will be encouraged in their own self-deception if we try to enlist them to support psyops and political warfare. The gambit will backfire.

We must step back and take a larger view in order to get our bearings and to reframe the issues so that neither we nor our allies fall victim to psyops by our opponents. We need the truth above all.

Let's reflect on the true nature of the threat to liberal democracy in the twentieth century.

The American theory of liberal democracy is based on the assumption that men are selfish, that they will fall into factional struggles, that they will try to use power for private ends. Our constitutional structure is a device for separating governmental powers and putting the law above any group or faction. Much is made of majority rule, but that is not the linchpin in our system. Democracies can be tyrannical if majorities abuse minorities. Minority and individual rights, in our system, depend on the law to restrain majorities.

In the early part of this century, Arthur Bentley, a distinguished student of political science, lost sight of that wisdom and advanced a new theory of pluralism. In his view, pluralism is competition among interests expressed by groups. Competition makes pluralism a reality and ensures it. Bentley dismissed institutional and constitutional approaches to political science, so popular in the nineteenth century, and in 1904 propounded group theory as an alternative in his book, *The Process of Government*.

In 1915, Bertrand Russell raised a serious issue for Bentley. What if one group does not play by the rules? What if a conspiratorial party decides to exploit a liberal democratic system to destroy it and monopolize the power? William Y. Eliot of Harvard made essentially the same criticism of Bentley in 1926. By that time, Eliot had the Bolshevik example as empirical support for Russell's theoretical apprehension.

Both Russell and Eliot insisted that the government, as an institution, is not just one more competing group, as Bentley argued, and as David Truman would suggest again in the 1950s with the revival of Bentley's interest group theory. Government has the duty to represent the state, to uphold the constitutional and legal basis for political competition, to stand above interest groups, to make groups play by the rules. Competing interests alone will not ensure liberal democracy. On the contrary, a radical group is soon likely to appear to upset the balance and destroy the system. By throwing out the concept of state as the rule keeper above interests, Bentley was opening the door to pluralism's enemies.

As we judge the political developments in the Soviet Union today, it is instructive to reflect on Lenin's destruction of the Constituent Assembly in 1918. The Constituent Assembly was an old idea among political reformers in Russia, and the Provisional Government had promised to hold elections for such a body in 1917. This body would create a new legitimacy, a new law-based state. The popularity of the idea was so great that even Lenin did not dare cancel the elections when he seized power. They proved to be the freest elections ever held in Russia, and the Bolsheviks, in spite of their manipulations, lost by a large margin. The Assembly was eventually allowed to convene, but Bolshevik tactics quickly destroyed it. Lenin was so worried about the power of this potential step toward constitutionalism in Russia that he personally supervised the Bolshevik troops and party cadres as they destroyed it. No law could be above the revolution, and the revolution was Bolshevik power in Lenin's meaning.

Here we have the barefaced example of the radical group variant destroying an effort to install a constitutionally based democracy, making all parties responsible before the law. We next saw it in Italy in the 1920s, then in Germany in the 1930s, a bit later in Japan. After World War II, we saw it in Yugoslavia, in all the states under Soviet control in East Europe, and in China in 1949.

For anyone who cares to look at the evidence, we now have the documents of the Bishop regime in Grenada. They read like a carbon copy of Lenin's tactics in 1917-21. All the behavioral evidence encourages the same conclusion about Nicaragua.

A number of Western scholars have tried to blame this menace to democracy on Stalin, relieving Lenin of the blame. Professor Cohen of Princeton University is one of the more ardent proponents of this view. He insists that the Soviet Union in the New Economic Policy (NEP) period was really pluralistic, an inchoate liberal development. It might not matter much whether Lenin or Stalin is to blame if it were not for some particularly wrongheaded ideas that thrive on Professor Cohen's analysis. NEP Russia was certainly politically fragmented. Competing groups indeed existed. Based on that evidence, he wants us to believe that had Lenin lived, or had Bukharin prevailed over Stalin, political pluralism would have evolved in the Soviet Union.

Like Bentley and Truman, he misunderstands what secures pluralism in Western democracy: the primacy of law over parties, even majority parties, law that protects minority rights. There was no such reality in Russia in the 1920s.

Many societies are fragmented politically, composed of warring groups, but they are not plural political systems. Lenin ensured that pluralism would not emerge in Russia when he destroyed the Constituent Assembly. It was the single hope to make Russia a state where law would prevail. Whether it would have succeeded can be debated, but Lenin's actions ensured failure.

As we watch Gorbachev's Soviet Union, we must not become confused about the threat to liberal democracy which brought U.S. power to the Eurasian rimland and has kept it there for four decades. It has not gone away. It has not even begun to recede. Is Gorbachev introducing constitutionalism, the primacy of law over Lenin's party? I see no convincing signs of it.

Consider what it would involve if he were. First, in the age of nationalism as the basis for state sovereignty, it would mean that he intends to let the national minority republics truly exercise self-determination. Perhaps Gorbachev will eventually do this, but he has not, and the evidence today indicates a nervous struggle in the Baltic states, Belorussia, and the Ukraine to restrain these developments.

Second, modern economic development depends on a key role for private property and market competition. The European Social Democrats, the original Marxist movement with which Lenin broke, have long since come to terms with private property and market economics. In doing so, they had to abandon the legitimacy of class struggle and revolution by violence. They have come to accept the primacy of bourgeois law, the necessity for constitutional legitimacy to protect property and individual rights. Notwithstanding all the talk under glasnost about economic reform and the need for market pricing, virtually no progress has been made in that direction under Gorbachev. Economic cooperatives are struggling, and they are still marginal compared to private farm plots as a market development. Soviet industry, where market pricing is most needed for significant economic progress, has yet to experience any shift to a market basis, the new law for the enterprise notwithstanding.

Third, neither legal reform nor market economics are going to succeed without dismantling the censorship system and permitting undisturbed lateral communications throughout Soviet society. As the Czechs discovered in 1968, a market economy depends on free information flows for advertising, for buying at the best price, and for protecting individual rights before the law.

Call it psyops, or political warfare, or any other name. Western scholars, intelligence officials, media elites, and political leaders in considerable numbers have been taken in by Soviet propaganda on these three critical measures of change. I do not know how calculated the Soviet effort is, but I have seen a constant stream of Soviet spokesmen drumming up support for Gorbachev and his policies in the West. They speak of a struggle, a war in Soviet politics between pro- and anti-reformers. They impress on their Western interlocutors the urgency of helping Gorbachev.

Perhaps a glimpse of their true objective is revealed in a point made by Zhurkin, Karagonov, and Kortunov in an article in December 1987 when they wrote:

> Unilateral measures, notwithstanding their large character, (the number of armed forces in 1955-58 alone was reduced by 37 percent), did not lead to a weakening of the USSR's international position. On the contrary, the second half of the 1950s through the beginning of the 1960s was time of rapid growth of the Soviet Union's foreign policy prestige and influence. . . . These measures did not lead to a weakening of security, or an increase in the threat of war. This is connected with the fact that they were accompanied by a wide-scale peace offensive, which hampered the mobilization of the West's efforts to place military pressure on our country.[1]

Is this not an open admission of a repetition of an old strategy? Did it not happen in the 1920s? Again during the Popular Front movement in the 1930s? Why is Soviet policy different today?

In a curious way, the Reagan administration managed to create a national strategy, including political warfare and psyops, in the early 1980s. It had actually begun to take shape in 1980 under President Carter. His sanctions in the aftermath of the Soviet invasion of Afghanistan were the most severe any president has taken since World War II. He also started the rebuilding of U.S. military power. Reagan continued and accelerated these policies, particularly the military buildup and the economic sanctions. He added a political and psychological dimension that Carter was loath to do. He called the Soviet Union an evil empire, and he proved willing to use military force against a Leninist regime in Grenada. He also stepped up support to anti-Soviet insurgencies, although the most important one was begun by Carter—support to the Afghan rebels.

President Ronald Reagan had his critics for this policy. The political left in Europe, even some of the center, reacted to characterizing the Soviet Union as an evil empire. This is very strange. Surely they did not believe it was a virtuous empire! Some Europeans lectured us on our lack of understanding *Realpolitik*. They told us that we did not know Russian history, that we did not understand Soviet politics. The ahistorical nature of this European outcry was surprising. It wholly ignored the poor historical record of Realpolitik for keeping the peace in Europe and the excellent record of peace under naive U.S. leadership. It ignored the fact that the best scholarship on Russian and Soviet history and politics has been done in the United States since World War II. The more I listened to these European voices, the more I wondered if it was not they who were parochial, regional, and ahistorical in matters of strategy and world politics. This is not true of most Europeans, but it does characterize a vociferous minority. I recall painfully listening to a German parliamentarian tell American guests at the German Embassy in Washington a few weeks after the Afghan invasion that President Jimmy Carter was turning a North-South issue into a East-West issue. I then realized how this German had lost his sense of geography and politics.

Within a few years after the evil empire speech, critics had to admit that Reagan's policies were working. Moscow stirred up a war scare, walked out of the arms talks, and hurt the peace movement in Europe. Later, when European political leaders stood firm on the INF deployments, Moscow changed course. The Soviet leadership reversed directions, came back to the arms talks, and accepted cuts that were earlier believed impossible for Moscow to consider seriously.

Unfortunately, President Reagan also seemed to lose his bearings under this new peace offensive. He almost fell into Gorbachev's denuclearization trap at Reykjavik, and Secretary Shultz's urge to summitry and media events gave Gorbachev the very openings he needed to get on the Western stage. What had become for a few years a *de facto* integrated U.S. political, military, and economic strategy began to collapse. We are seeing the results. U.S. policy has been reactive and without focus. The public consensus behind the early Reagan policy has now been displaced by confusion. "Is there still a Soviet threat?" they are asking.

The Soviet strategy has changed, but the fundamental character of the threat has yet to alter. Gorbachev has renounced neither Leninism nor socialism.

THE PRESENT NATURE OF SOVIET POLITICS

You may justly ask if I am not ignoring a lot of things now occurring in the Soviet Union. The answer is no. I see them differently from popular interpretations.

Mr. Andropov inherited from Brezhnev a corrupt and decadent party bureaucracy. It had not been purged since Khrushchev's days. It had learned how to prevent the traditional purge through an exchange of party documents and a reduction in membership, creating room for new, more vigorous members and promotion for the more aggressive members. Andropov tried a new tactic, attacking alcoholism and corruption, but he died too early to carry through. Gorbachev, who probably owes his rise to power more to Andropov than anyone else, has taken up where Andropov failed.

His tactic is an unusual yet promising one for overcoming bureaucratic resistance. Through glasnost, he has enlisted the intelligentsia to criticize the bureaucracy, to create an atmosphere justifying large turnovers in cadres. In so doing, he has had to permit a variety of voices to be heard, many seeking fundamental change in the Soviet system. Liberals, Russian nationalists, neo-Stalinists, anti-Semites, circles which long for an idealized peasant society, national movements, and others have been allowed to argue their cases. Perestroika is Gorbachev's response. But precisely what is perestroika? To date, it has been mainly cadre turnover. The Politburo and the Secretariat are virtually filled with new faces. Turnover in the regional party organizations is mixed. In some places it has been significant. In others, especially the Ukraine, it has been virtually nil. "Why are new *apparatchiki* better than old ones?" you may ask. Gorbachev's answer is clear: They are "new thinking" apparatchiki.

New thinking, if it has meaning other than loyalty to Gorbachev, is willingness to be more responsive to central direction at the expense of local and ministerial interests. A simple revitalization of the party is not enough. Gorbachev realizes that the international correlation of forces is against the USSR. To reverse those trends, he must make some major policy and resource shifts. He must break the corruption in the economic apparatus. He must shift resources away from military outlays in order to bring the industrial base up to a level that will permit a continuation of the East-West competition. He must have Western cooperation, and he knows he must pay a significant price for that assistance.

The Brezhnev gains in the Third World have not paid off as well as was anticipated. On that front, Gorbachev is resorting to the military principle of economy of force. He is withdrawing in some areas, consolidating in others, and using cheaper means, mainly political, to deny Western gains in the face of his retreat.

On the front with the industrial West, he has placed the main emphasis. He is attacking so-called anti-Soviet forces head-on, inviting them to Moscow, wooing them, offering economic opportunities, and allaying their fears about security. As Zhurkin and his fellow authors point out, he is using the peace offensive to cover his reduction in military outlays.

On the Chinese front, he has made major concessions to two of the three Chinese demands, Afghanistan and Cambodia. He can be expected to make concessions on troops on the border, and his withdrawals from Mongolia are a step in this direction.

What has caused this strategic redirection of Soviet policy? Moscow lost the war in Afghanistan. It won the quantitative arms race, but it is losing the qualitative arms race. Colonel General Moiseyev, in his first major article since becoming chief of staff, argues that the Soviet Union must avoid a qualitative arms race with the West. Moscow has also been losing the political competition in the Third World while paying large costs to support client states. Finally, the Soviet Union has been largely cut off from the high technology of the Western economies. Poor internal performance economically coupled with external factors explains the new Soviet strategy.

It may be objected that while all these things are true, it does not mean that we are not seeing fundamental change in the Soviet Union; that the Cold War is really over; and that, whether he likes it or not, Gorbachev is presiding over forces that will ensure the end of Leninism as we have known it. Certainly some of the Soviet intelligentsia seem to be dedicated to this outcome. Perhaps they will succeed, but it is far too early to pronounce them victorious. We should not rule out this outcome, and we should certainly watch for evidence that it is becoming more likely.

WHAT SHOULD BE OUR RESPONSE TO GORBACHEV'S OFFENSIVE?

This point raises the question of what we should do about Gorbachev's policies. That is easy to decide, but not easy to implement.

Clearly we should keep up the military qualitative competition and push the Soviets in regional competition. In the economic area, we should not deny assistance but rather make it contingent on seeing the Soviet economy move to market pricing in at least half of its industrial sector. To achieve that, Gorbachev would have to introduce private property and genuine decentralization of the economy. The concomitant political decentralization might well cost the Communist party its monopoly. In the event, we could safely begin to think about the end of the Cold War and engage in economic interaction. On the political and diplomatic front, we must devise initiatives that truly test the Soviet peace offensive. We cannot simply call perestroika a sham. We must advance policies that tend to unmask it, if it is a sham.

Soviet power is in decline, but it has not become benign. Let me give an example of its strength. In fragmented societies, the Leninism concept of organizing institutions to maintain control over dissident sectors and groups is attractive. Our concept of liberal democratic institutions is not. Our recipe of elections and representative government is too often a formula for losing control, for falling prey to the radical group, for feckless responses to insurgencies. Psyops aimed at winning hearts and minds are usually naive politically in these situations.

Even if Moscow reduces its commitments to national liberation movements in the Third World, they will not necessarily stop. In Latin America and the Philippines, they will look for other sponsors and continue on their own. The bacillus of Bolshevism will not always die without Soviet support.

This raises the issue of noncommunist threats to democracy. Indeed they exist. Right-wing dictatorships are just as unacceptable to our democratic ideals as communism. I tend to agree with those who argue that right-wing dictatorships are less of a threat because they do tend to evolve to become democratic systems while communist systems do not. It makes sense in our tactics to make temporary alliances with these regimes against the communist threat in some cases. The danger, however, always exists that we become the prisoners of the policies of these regimes and bear the onus of their dictatorial policies.

There is no easy solution to this dilemma, but we should make it clear that when the communist danger is removed, we intend to attack the right-wing authoritarians. I believe it is possible to educate the U.S. public to understand such a strategy. Until we do, we are vulnerable to the Soviet psyops that mobilize Western groups against U.S. policy in places like Latin America.

NOTE

1. V.V. Zhurkin, S.A. Karagonov, and A.V. Kortunov, "On Reasonable Sufficiency," *USA: Economics, Politics, and Ideology* 12 (1988): 14-20.

3

U.S. Military Psychological Operations: Past, Present, and Future

Alfred H. Paddock, Jr.

Psychological operations are broadly defined as the planned use of communications to influence human attitudes and behavior. They consist of political, military, and ideological actions conducted on target groups to create desired behavior, emotions, and attitudes. If used properly, psyops will normally precede, accompany, and follow all applications of force. This course of action should be carried out under the broader umbrella of national policy, and the military component of the overall psychological operations effort should be coordinated fully and carefully with other agencies of government.

More specifically, psyops can be used to demoralize, disorient, and confuse hostile groups. Against such groups, psyops are employed as an offensive weapon to enhance the overall effectiveness of military operations. They can also be used, however, to unite, inform, and bolster the morale of nonhostile groups. When targeting neutral or friendly groups, they are used to support military objectives by developing cooperative attitudes and behavior in the targeted group.

Any discussion of U.S. psychological operations and political warfare must include a consideration of U.S. military psychological operations (psyops) capabilities. This chapter provides a brief perspective on the evolution of U.S. military psyops during the twentieth century, examines the conditions that led to a revitalization of these capabilities during the mid-1980s, contrasts this with the international and domestic environment faced by the U.S. military today, then outlines some of the major challenges facing U.S. military psyops in the future.

EVOLUTION OF U.S. MILITARY PSYCHOLOGICAL OPERATIONS

The level of interest in military psychological operations during this century has been episodic, basically rising and falling during and after the major conflicts in which U.S. forces have been committed. Over this period, most of the activity in military psyops centered on the Army. The following brief historical perspective, therefore, will focus on the Army's experience as illustrative of the fortunes of military psyops.

While giving psychological warfare only token recognition in World War I, the Army established the Psychological Warfare Sub-Section of G-2 in the War Department, and the Propaganda Section, G-2, General Headquarters, American Expeditionary Forces. Military tactical psychological warfare centered on the production of leaflets; radios did not exist as a means of communication and loudspeakers were primitive. Military propaganda concentrated on producing surrender appeals; balloons and airplanes were the primary method for their dissemination.

From 1918 to 1941 no psychological warfare office existed at the War Department. The lessons of experience were lost, and by 1941 only one officer on the War Department staff had psychological warfare experience in the previous war.

During World War II, most of the Army's operational work in psychological warfare took place at the theater level, where the responsible organization was normally designated a Psychological Warfare Branch (PWB). The largest of these, the PWB at Allied Forces Headquarters, was activated in North Africa in November 1942 at the order of General Dwight D. Eisenhower, and then expanded in February 1944 to the Psychological Warfare Division, Supreme Headquarters, Allied Expeditionary Force (PWD/SHAEF). PWD/SHAEF defined psychological warfare as "the dissemination of propaganda designed to undermine the enemy's will to resist, demoralize his forces and sustain the morale of our supporters."

The basic army field operating unit for psychological warfare was the Mobile Radio Broadcasting (MRB) company. The equipment for these units was unlike anything conventional soldiers had seen before in the field—public address systems, radios, monitoring sets, loudspeakers, typewriters, mobile printing presses, and leaflet bombs. MRB units were usually divided by the separate army groups and field armies into small teams, often to work in direct support of frontline conventional combat units. Five such companies eventually served under PWD/SHAEF. Although these units were the result of improvisation in 1943 and 1944, the doctrinal and organizational concepts they embodied reappeared in the psychological warfare units formed during the Korean Conflict.

During 1945-46, army psychological warfare staffs and units dissipated with the general demobilization of the military establishment. Despite the efforts of a few senior civilian and military officials to retain a military psyops capability, when the North Koreans attacked South Korea in June 1950, the Tactical Information Detachment—organized at Fort Riley, Kansas, in 1947—was the only operational psychological warfare troop unit in the U.S. Army. Sent to Korea in the fall of 1950, it was reorganized as the 1st Loudspeaker and Leaflet (L&L) Company and served as the 8th Army's tactical propaganda unit throughout the conflict. Tactical propaganda, sometimes called combat propaganda, was directed at a specific audience in the forward battle areas and used in support of localized operations. Mobile loudspeakers mounted on vehicles and aircraft became a primary means of conducting tactical propaganda in Korea.

To conduct full-scale strategic operations, the 1st Radio Broadcasting and Leaflet (RB&L) Group was organized at Fort Riley and shipped to Korea in July 1951. The 1st RB&L Group was specifically designed to propagate strategic propaganda in direct support of military operations. This propaganda was intended to further long-term strategic aims and was directed at enemy forces, populations, and enemy-occupied areas. The 1st RB&L Group had the equipment and capability to produce newspapers and leaflets, and to augment or replace other means of broadcasting radio propaganda. The Group supervised a radio station network, known as the Voice of the United Nations, and often produced more than 200 million propaganda leaflets a week that were disseminated by aircraft or by specially designed artillery shells. The leaflets expressed various themes. Some, for example, offered inducements for enemy soldiers to surrender; others were intended to bolster the morale of Korean civilians by proclaiming U.N. support.

Although the RB&L Group was a concept accelerated to meet the requirements of the Korean conflict, it performed functions similar to those used in psychological warfare in World War II. Its Mobile Radio Broadcasting (MRB) Company bore a direct ancestral linkage to the mobile radio broadcasting companies formed under PWD/SHAEF to conduct propaganda operations in North Africa and the European theater during 1944-45. However, the MRB companies during World War II were organized to perform tactical psychological warfare; radio later became an essentially strategic weapon that had no place in a purely tactical psychological unit. Both the strategic propaganda concept embodied in the RB&L group and the tactical propaganda idea expressed by the L&L Company were to figure prominently in the psychological warfare capability subsequently formed as part of the Psychological Warfare Center in 1952.

As originally established at Fort Bragg, North Carolina, the Psychological Warfare Center consisted of a Psychological Warfare School, the 6th Radio and Broadcasting Group, a Psychological Warfare Board, and the 10th Special Forces Group. The mission of this unprecedented center was:

> To conduct individual training and supervise unit training

> in Psychological Warfare and Special Forces operations; to develop and test Psychological Warfare and Special Forces doctrine, procedures, tactics, and techniques; to test and evaluate equipment employed in Psychological Warfare and Special Forces Operations.

After an initial burst of activity fueled by the Korean Conflict and by fears of a possible outbreak of war in Europe, interest in the Psychological Warfare Center began to dissipate. In 1956 its title changed to the Special Warfare Center, but by the early 1960s the Army's psychological operations capability had eroded.

Consequently, an insufficient base of psyops-trained officers existed to call upon when the 6th psyops battalion was activated in Vietnam in 1965. By 1967 the Army's psyops forces in Vietnam had been expanded to a group (the 4th) with four battalions, one in each of the four Corps Tactical Zones (CTZ). The group was under the control of the Commander, U.S. Military Assistance Command, Vietnam, with the J-3's Psychological Operations Division exercising direct staff supervision. The Joint U.S. Public Affairs Office provided U.S. psyops policy guidance not only to the civilian agencies but also to all military psyops elements.

In addition to providing tactical support to field force commanders, the 4th psyops Group assisted the South Vietnamese government in its communication effort down to the hamlet level. The group headquarters operated a 50,000-watt radio station and high-speed heavy printing presses, published a magazine for Vietnamese employees working for U.S. government and civilian agencies, and had a capability for researching and developing propaganda materials.

Psyops battalions had light printing presses, a research and propaganda development capability, personnel to work with the U.S. Air Force Special Operations units for aerial leaflet and loudspeaker missions, and ground loudspeaker and audiovisual teams. Loudspeaker and audiovisual teams operated with U.S. divisions and brigades or with province advisory teams. The 7th psyops Group in Okinawa provided valuable backup support in printing and high-altitude leaflet dissemination.

Thus, during the height of U.S. involvement in Southeast Asia, the Army stationed psyops units at Fort Bragg, North Carolina, and in Germany, Panama, and Okinawa, in addition to the 4th psyops Group in the Republic of Vietnam. By the mid-1970s, however, all that remained in the active component was an understrength, overcommitted, antiquatedly equipped group at Fort Bragg—a condition that did not improve significantly for ten years.

1980s REVITALIZATION OF PSYOPS

The mid-1980s saw an upturn in the fortunes of military psyops, made possible by several conditions. First, President Reagan came into office with a strong anticommunist orientation, characterized the Soviet Union as the "evil empire" early in his tenure, and directed a program of support to noncommunist insurgencies (the "Reagan Doctrine") around the world. With public opinion supporting him, President Reagan embarked upon a large buildup of the U.S. defense establishment, which also benefitted military psyops.

Next, the president issued a number of national security directives which outlined his approach to the psychological dimension of national security. His initial national security strategy, announced in the summer of 1981, contained four basic components: diplomatic, economic, military, and *informational.* This emphasis on the psychological component was retained in subsequent national security strategies.

In January 1983, President Reagan signed National Security Decision 77, entitled "Management of Public Diplomacy Relative to National Security." This directive defined public diplomacy rather broadly, stating that it "is comprised of those actions of the U.S. Government designed to generate support for our national security objectives." The term generally has evolved to embrace a broad range of informational and cultural activities. The decision also established an interagency mechanism to plan and coordinate public affairs, information, political, and broadcasting activities of the U.S. government.

Finally, in early 1984 the president directed the Department of Defense (DOD) to rebuild its military psychological operations capabilities. In response to this directive, the secretary of Defense launched a major evaluation of the department's capabilities and needs in the area of psychological operations. This evaluation concluded that military psychological operations capabilities had been allowed to atrophy over the past decade—a conclusion reached by many in the psyops community well before this time, but not by senior policymakers. Across-the-board deficiencies had developed in policy guidance, roles and missions, doctrine, organization, force structure, operational concepts, planning, programming, training, logistics, intelligence support, readiness, personnel programs, and—most importantly—attitude, underscoring the need for education and heightened awareness at all levels of military and civilian organization.

DOD Psyops Master Plan

The secretary of Defense elected to use a DOD psyops Master Plan as the framework for rebuilding military psyops capabilities. Approved by Secretary Weinberger in mid-1985, the plan serves as a comprehensive design for the fundamental improvement of the department's capabilities to employ psychological operations effectively, worldwide, in support of national objectives, in peace and crisis, and at all levels of conflict.

Any plan for revitalizing psyops must address two broad requirements: how to develop the capability for war, and how to prepare for contingencies short of war. The requirement to prepare for war—and the transition from peace to conflict—must remain our first responsibility, but the challenge for the Department of Defense to contribute to our nation's peacetime "war of ideas" with the Soviet Union must not be neglected. The Master Plan addresses these two requirements.

The plan specified a number of remedial actions—over two hundred, in fact—to be implemented over several years. There were a few essential themes.

The first theme was the need to develop a *comprehensive joint doctrine* for the formulation, direction, coordination, and conduct of psyops in peace, crisis, and war. In effect, this doctrine should provide the foundation for the revitalization effort. Among other things, it should enunciate the function of psyops as a force multiplier in all types of military activity, establish the conceptual framework for planning and implementation, and delineate roles and responsibilities of the several components. The Joint Chiefs of Staff published this doctrine in February 1987.

The development of doctrine must be paralleled by *major improvements in psyops planning*—another essential theme of the Master Plan. There is insufficient human talent devoted to full-time, meaningful, sustained psyops planning at appropriate staff levels. We have seen some evidence of progress on this major deficiency. Creation of a psychological operations directorate, the first such office to exist in the Office of the Secretary of Defense (OSD) in over 20 years, was an indication of the seriousness with which Secretary Weinberger undertook this revitalization effort. The psyops staff element in the Office of the Joint Chiefs of Staff (OJCS) was upgraded from a branch to a division, and increased in personnel. The Department of the Army staff created a psyops and civil affairs division, where before only one officer was devoted full-time to this activity. Among the unified commands, USSOUTHCOM created a psyops detachment to augment its staff capability.

Much more remains to be done in this area, however. Perhaps the key requirement is that staff officers be trained formally in this specialized area, and that psychological operations be an integral part of the operational course of action in any plan. A few innovative and resourceful staff officers can make a vast difference, if positioned so as to have access to a command's major planning activities. The new Joint psyops Staff Planning Course developed by the Army should go a long way toward helping to provide the trained personnel required in all services.

Development of adequate numbers of psyops planners is an area where the Air Force, Navy, and Marine Corps can make an important contribution to the secretary of Defense's revitalization effort. The conduct of psychological operations is not the exclusive domain of specialized units such as those in the Army. Over the near term, it is not likely that the other services will opt to field new psyops units. On the other hand, there is every reason to expect those services to develop fully qualified psyops planners to meet their own needs and to provide their proportionate share of planners on joint staffs.

A more sophisticated psyops planning capability would allow the services to more effectively employ "psychological actions," those actions carried out primarily for their psychological effect—sometimes called "propaganda of the deed." These are activities that do not normally involve the use of printed or audiovisual media to disseminate the propaganda message, but which are planned to have a psychological impact. Exploitation of military exercises, deployment of military forces, contact with foreign nationals through port visits and civic action, and specific types of combat operations are examples of psychological actions that are possible with properly trained psyops planners to advise military commanders.

Closely related is the need to *educate our officer corps on psychological operations*, a third theme. As indicated earlier, the root cause of the atrophy of our military psyops capabilities has been lack of understanding of psychological operations, their value, and their application. Some improvement has been seen in this critical area as a result of frequent briefings of senior commanders and staff officers by psyops personnel, the professionalism of psyops units in contingency planning and support of conventional forces on joint training exercises, and the steady improvement in quality of psyops studies and assessments in support of the unified commands and national-level agencies (the latter aided considerably by the increased hiring of high-quality civilian intelligence analysts).

Certainly the psyops courses currently being presented by both the Air Force and the Army will help address the deficiencies in this area of psyops awareness and understanding. The Air Force's new Joint Senior Psychological Operations Course, conducted four times a year by their Special Operations School at Hurlburt Field, Florida, provides selected senior officers and civilians with an awareness of how psychological operations can support U.S. national objectives throughout the spectrum of conflict. This short course, in particular, shows considerable promise as an educational tool.

But, as was the case before the Vietnam conflict, psyops instruction in our mainstream service school system—where our future commanders and staff officers are trained—is limited or nonexistent. Its absence not only makes the psyops community's job more difficult in educating supported units on the capabilities and limitations of this unique weapons system, it also quite naturally has a negative effect when priorities concerning force modernization are being set.

To address the loss of psyops institutional memory, the under secretary of Defense for Policy sent letters to commandants of the senior service (the war colleges) and command and staff colleges emphasizing the secretary of Defense's intention to revitalize our military psyops capabilities and offering a presentation on psychological operations by OSD. This presentation is intended to stimulate interest in subsequent incorporation of more extensive treatment of psyops in service school curricula. To accomplish this, the military services have been asked to review all levels of their officer and enlisted professional military education (PME) curricula to develop new educational goals which will upgrade their psyops PME. As a second part of this approach, OSD has initiated a contract to develop applicable psyops curriculum material for the services. This is just a start—but an important one, nonetheless—if understanding of psyops is to be institutionalized for the long haul.

The next broad theme encompasses the need to *modernize our psyops force structure*, both in terms of personnel and equipment. Not surprisingly, a considerable shortfall in psyops forces exists. The Navy has a radio and television production capability in its reserves which is very good, plus a 10 KW mobile radio transmitter assigned to its Tactical Deception Group (Atlantic) which can be used to support psychological operations activities. The Air Force has a National Guard Squadron of specially fitted C-130 aircraft for radio and TV support of psychological operations, as well as other duties; it also has a handful of officers with psyops expertise serving in key positions in the Pentagon, among the unified commands and at their Special Operations School at Hurlburt Field, Florida. The Marine Corps has two civil affairs groups in its reserves with psyops as a secondary mission. Only the Army, however, has active duty forces dedicated solely to psychological operations.

The 4th Psychological Operations Group at Fort Bragg, North Carolina, is what remains of the Army's active psyops capability following the United States' withdrawal from Vietnam. Today, its missions and responsibilities are many and worldwide in nature. The group provides support to all levels of the Department of Defense, from the unified command through the division. It provides support to both conventional forces and special operations forces. In addition, it is often called upon to provide support directly to national level agencies and organizations. If any one military unit can be adjudged a "national asset," surely the 4th Psychological Operations Group fits the requirements.

Essentially, a military psychological operations unit engages in two broad categories of activity: research and analysis and operations. The first activity consists of continuous monitoring and assessing of the psychological environment in specific foreign nations to determine how that environment affects the formulations and execution of U.S. policies and actions. This research and analysis results in the publication of studies and assessments that are unique. They provide the foundation for establishment of psychological objectives to support U.S. goals related to foreign nations or groups. Research and analysis is therefore essential to the accomplishment of the second broad category of activity: the planning and executing of specific psychological operations campaigns which employ communications media and other techniques to cause selected foreign groups and individuals to behave in ways that support U.S. national and military objectives.

Thus, a military psyops unit in peacetime conducts research and analysis of specific geographic regions and target audiences, develops psyops plans to support conventional and special operations units, and participates in field exercises that employ these plans. Because of the paucity of psyops expertise at unified commands, the 4th Group also provides staff assistance and advice to those headquarters and to other major commands.

It should be eminently clear from the foregoing that one active duty psyops organization, consisting of a group headquarters, a strategic dissemination company with printing, radio and television production capabilities, and four regionally oriented battalions, is insufficient to support all unified command requirements in mid- or high-intensity conflict. The reserves are therefore a vital component of the "psyops community"; fully 80 percent of the Army's psyops mobilization capability lies in its reserve component units. Serving as the Army's planning agent under the CAPSTONE program (which links reserve units with the units they would support upon mobilization) the 4th Group coordinates the wartime planning efforts of reserve units and provides training assistance.

Generally speaking, then, the active component 4th psyops Group acts as a "strategic nucleus" for the psyops community; it provides the bulk of peacetime research and analysis support, responds to peacetime and low-intensity conflict requirements, provides direction and guidance to the psyops community for wartime planning and participation in peacetime exercises, and provides the active component command and control nucleus for general or partial mobilization of reserve component forces. The reserve component performs its planning and training responsibilities under the CAPSTONE program and prepares for general or partial mobilization in support of the unified commands.

Paradoxically, the successful CAPSTONE program underscores one of the psyops community's most glaring weaknesses: its limited capability to respond to peacetime and low-intensity conflict requirements. As has been stated, for mid- or high-intensity conflict requirements, either partial or general mobilization of the reserve component is required. Conversely, the active component must be relied upon for almost all peacetime and low-intensity conflict requirements. The most probable demands on psyops resources in this environment would be for support of DOD and non-DOD agencies, staff assistance to unified commands, unscheduled studies and assessments oriented to crisis areas, and advisory mobile training teams for the military forces of friendly Third World nations. These demands, in addition to the vital task of continuing to plan and train for mid- and high-intensity contingencies, will strain to the utmost the active component of 4th psyops Group.

The Army has made progress in bringing the 4th psyops Group up to its authorized strength, after many years of neglect in this area. Even at full strength, however, this is an inadequate active component force posture, particularly in view of increasing peacetime requirements being placed on the 4th by the unified commands and the requirement to support vital national-level undertakings.

Realistically, the reserve component will continue to provide the bulk of our psyops capability for mid- or high-intensity conflict requirements. In response to a psyops Master Plan tasking, the Army conducted a major study to determine the number of additional reserve component units needed to meet these contingency requirements. Approved by the Chief of Staff, Army, in October 1986, the action will add over 2700 spaces to our RC mobilization capability in the 1990s.

In terms of the quality of psyops personnel, the Army's development of a psyops military occupational specialty (MOS) for enlisted personnel is a welcome accomplishment. Initial reports of the high quality of personnel being trained under this new specialty are encouraging. On the other hand, the Army's decision to remove its officer psyops and civil affairs MOS from the foreign area officer (FAO) specialty is disturbing.

This change is disturbing because it separates psychological operations from the foreign area officer specialty that provides its intellectual lifeblood. The core of the area expertise, knowledge of foreign cultures, and analytic capability of psychological operations is in the FAO specialty. Like intelligence, the strength of psychological operations is in its people.

The Army's initial decision was to include both psyops and civil affairs in the special operations forces functional area. More recently, however, the Army has developed a separate functional area for psyops and civil affairs that replicates much of the training and professional development of the FAO program. It remains to be seen, however, whether this initiative will provide the quality officers for psyops previously produced by the FAO program.

With respect to modernization of psyops equipment, there are many encouraging initiatives under way in the Army. The 4th psyops Group's new media production center represents a quantum leap in military psyops capability to operate in a modern audiovisual communications environment. A number of other promising projects to upgrade active and reserve component print, radio, loudspeaker, and audiovisual capabilities are in various stages of progress and must be followed up vigorously. Similarly, the Air Force has allocated funds to modernize its National Guard aircraft dedicated to support of psyops.

The reserve components, however, lack some essential items of psyops equipment, particularly loudspeakers. All of the services need to establish a funded program for psyops-related research and development. The Air Force, for example, needs a device that can rapidly release leaflets in high altitude operations. A vigorous interservice development and acquisition program aimed at the likely environment of the 1990s and taking advantage of the ready availability of state-of-the-art equipment in the commercial market would be a step in the right direction.

One of the more controversial themes of the Master Plan was the *organizational separation of psychological operations from special operations*. In general, the subordination of psyops elements to special operations detracted from recognition of the overall applicability of psychological operations in times of peace, crisis, and war. The Master Plan, therefore, called for the separation of psychological operations from special operations throughout the Defense Department, including at departmental level and all headquarters and staffs of the Joint Chiefs of Staff, the services, the unified and specified commands, and component and subordinate commands. The Master Plan outlined several reasons favoring separation.

Planning, particularly in the unified and specified commands, suffered because the sole psyops staff planner is usually located in the special operations staff element and is only employed part-time in psychological operations. This subordination detracted from the broader responsibility of planning psychological operations support for the theater's total requirements, particularly in those missions that link military psychological operations and national objectives, policy, and strategy, and require in-theater interagency cooperation.

Further, the subordination of psychological operations units to the special operations field, including its command and staff structure, contributed to the lack of understanding of psychological operations and its uses and capabilities by military officers and senior civilian and military leaders. This association and subordination caused many to conclude that psychological operations are focused primarily in support of special operations missions.

While psyops do have a mission in support of special operations, both in high-intensity conflicts and in low-intensity operations, they also have a much broader application in peacetime and crisis, with or without accompanying military operations, and across the entire spectrum of conflict. Only 10 percent of the Army's psychological operations force, active and reserve, are designated by current contingency plans to support special operations forces in wartime.

In this connection, the argument for separation of psyops from special operations is implicit in the Army's and Joint Chiefs of Staff's arrangements for wartime command and control of psychological operations. Most of the psychological operations forces in wartime are aligned with a chain of command in the unified commands that is totally separate from special operations forces. Psyops units are *combat support* forces and are employed at both strategic and tactical levels from the theater to the division as a matter of routine; special operations forces are employed primarily as strategic assets on an exceptional basis.

Subordination of psychological operations to special operations could cause military psychological operations forces to focus their very limited resources in support of special operations, to the detriment, in particular, of the increased peacetime role for military psyops. For all of these reasons, the Master Plan indicated that increased understanding and employment of psychological operations lies with missions other than special operations.

Because of its controversial nature, the separation directed by the Master Plan was slow in implementation. To be sure, some important initial steps were taken. The new psyops directorate in OSD was separated from special operations, in terms of policy responsibilities. Psyops were removed from the Joint Special Operations Agency and placed under the J-3 on the Joint staff. An important precedent has been established on the Department of the Army staff by the creation of a psyops and civil affairs division separated from special operations. And two unified commands—USSOUTHCOM and USCENTCOM—have effected the separation within their staffs.

But for the large part, separation did not take place outside the Pentagon among the unified commands, subordinate service headquarters, and at the operational level. A major reason for this delay, of course, was the uncertainty resulting from the congressionally mandated reorganization of special operations forces. This subject will be addressed in more detail below.

The separation issue was closely related to the final major theme of the Master Plan: *creation of a Joint Psychological Operations Center*. According to the Master Plan, psychological operations are sufficiently important to warrant the creation of a separate center dedicated to the long-term development and nurturing of this unique capability. This center would become the organizational and intellectual font of psychological operations within the Department of Defense.

Among the responsibilities for the center envisaged in the Master Plan were long-range, strategic psychological operations plans; doctrine and operational concepts; continuing education and training of personnel; research and analytical studies; and development of equipment. The center was also to assist the OJCS and OSD to develop, plan, and coordinate the defense portion of national psychological operations activities and to assist the unified and specified commands in their planning of psychological operations.

The structure of the Joint Psychological Operations Center was to consist of two separate but mutually supporting elements, one operational and the other developmental, each indispensable to the other. These elements were to have been the 4th Psychological Operations Group and psyops spaces in the Army's John F. Kennedy Special Warfare Center and School. This concept therefore would have used existing psychological operations units and personnel spaces to provide the nucleus of initial manpower requirements, an important and realistic factor in the present resource-constrained environment.

The Master Plan indicated that initially the center probably would have to be under army management because the other services would have little to contribute to manpower. However, as representation from the other services increased to more than token level, the joint character of the center was to be emphasized by making its command rotational among the services. As with the separation theme, however, implementation of the center has been held in abeyance pending resolution of several issues resulting from the congressionally mandated reorganization of special operations forces.

In summary, these were the major themes of the secretary of Defense's Master Plan for the revitalization of military psychological operations: formulation of comprehensive joint doctrine, indoctrination of the officer corps, improvement of staff planning, modernization of the force, separation of psyops from special operations, and creation of a joint psyops center.

Special Operations Force Reorganization

Obviously this discussion of the Master Plan must include the congressionally mandated reorganization of special operations forces. In mid-October 1986, Congress passed a Special Operations Force (SOF) reform package sponsored by Senators William Cohen (R-Maine) and Sam Nunn (D-Georgia). The law created a unified SOF Command under a four-star general or flag officer, an assistant secretary of Defense for Special Operations and Low-Intensity Conflict, a board for Low-Intensity Conflict within the NSC, and a deputy assistant to the president for National Security Affairs for Low-Intensity Conflict. Particularly important, it directed the secretary of Defense to create for SOF a major force program category for the Five-Year Defense Plan of the Department of Defense.

After the passage of this legislation, a major question arose on whether or not psyops and civil affairs units should be included in the assignment of forces to the new U.S. Special Operations Command (USSOCOM). There were good arguments on both sides of this issue. And, the secretary of Defense's previous guidance to separate psychological operations from special operations—as part of his decision concerning implementation of the 1985 DOD Psychological Operations Master Plan—had to be taken into consideration. After a lengthy review of this issue, Secretary Weinberger decided to assign Army and Air Force active and reserve component psychological operations and civil affairs units to USSOCOM effective October 15, 1987. At the same time, the secretary recognized that this decision affected, in part, his earlier guidance in the Psychological Operations Master Plan. He therefore directed that the Master Plan be reevaluated and modified as necessary to take into consideration these changed circumstances. This revalidation process is currently nearing completion.

Thus, the revitalization of military psyops capabilities which began to take shape during the mid-to-late-1980s was fueled by a presidentially inspired public perception of the common threat (the Soviet Union), a "defense consensus" which enabled massive increases in military budgets, several key Reagan administration national security directives which provided impetus for the DOD, and the "top-down" emphasis of Secretary of Defense Weinberger and a few key advisers which resulted in development of the DOD psyops Master Plan.

THE CURRENT MILITARY PSYCHOLOGICAL OPERATIONS ENVIRONMENT

A comparison of these earlier conditions with the present environment provides some rather stark contrasts. The most dramatic change has occurred in the public's perception of the threat. Americans are more interested today in negotiations with the Soviet Union than confrontation. Clearly, an important stimulus for this change in attitude has been the emergence of Mikhail Gorbachev as Soviet leader. His domestic political and economic reforms, withdrawal of troops from Afghanistan, cessation of jamming Voice of America broadcasts, announced intention to reduce Soviet military forces by 500,000 men and cut their budget by 14.2 percent, and the INF arms control agreement he negotiated with President Reagan are just some of the actions that have contributed to a lessening of international tensions. Mr. Gorbachev's policies led Margaret Thatcher to conclude in November 1988, that the Cold War had come to an end. This conclusion by a staunch anti-Communist was reflected in polling data of U.S. public attitudes, which depict a startling change in just over a two-year period. By July 1988, 94 percent of voters believed that U.S.-USSR relations were stable or getting better.

This lessening of the perception of the threat, coupled with concerns over the burgeoning federal budget deficit, has led to domestic and congressional pressures for reductions of military spending. Responding to these pressures, President Bush has already announced that he will trim President Reagan's proposal for defense budget increases in fiscal year 1990. This "zero-real-growth" proposal is only a harbinger of considerably tighter defense budgets over the next several years, a prospect which will force some tough decisions within DOD on military spending priorities—to include rethinking the size of our force structure. It is clear that the American public no longer supports continued unrestricted growth of military spending. The "defense consensus" has vanished.

And so have the Reagan national security directives that provided the vital impetus for DOD to revitalize military psyops. As is the case with every new administration, President Bush and his national security advisor, Brent Scowcroft, have announced their intention to conduct a thorough review of U.S. national security strategy. In the meantime, the fate of Reagan administration directives is uncertain; there is no indication yet to what extent those pertaining to the psychological dimension will be incorporated as part of President Bush's national security strategy. Finally, whereas before Secretary Weinberger provided "top-down" emphasis to revitalize military psychological operations, at present the Department of Defense is in transition.

CHALLENGES FOR FUTURE MILITARY PSYCHOLOGICAL OPERATIONS

Thus the challenges for military psychological operations in the future are formidable. Caution is appropriate when contemplating Mr. Gorbachev's policies. His internal difficulties are considerable, the military reductions he has announced have not yet taken place, and there is always the possibility that he could be replaced by a hardliner on U.S.-USSR relations. Because we cannot foretell the future, U.S. strategic policy should be guided by a realistic evaluation of Soviet military capabilities, which remain formidable. Nevertheless, clearly the U.S. public perception of the threat has changed dramatically, and military psyops advocates must adjust to this condition.

They must do so by demonstrating the applicability of psychological operations in all environments, to include support of special operations forces, low-intensity conflict, and conventional forces. A strong case should be made that psyops play important roles in enhancing deterrence. Similarly, military psyops must continue to demonstrate that they can play a valuable supporting role to the peacetime activities of other government agencies, such as the USIA and the CIA. Military psyops can—and should—continue to support programs of the National Command Authority, when appropriate.

With regard to the coming period of more stringent defense budgets, the challenge for military psyops is to fight for their proportionate share of resources—or, put another way, to ensure that they do not suffer a disproportionate share of reductions. Being assigned to the USSOCOM, which recently acquired its own budget, should help in this endeavor. This means that for the first time special operations forces have fiscal independence from the services in forming a five-year defense spending plan.

Nevertheless, military psyops must demonstrate a fiscally responsible attitude by making maximum use of existing resources. This includes not only continued reliance on the psyops reserve components for mobilization in high- and mid-intensity conflict, but also investigating ways in which specially trained reservists can be used to augment active component units during peacetime and low-intensity conflict requirements. Military psyops must avoid the establishment of redundant capabilities, and should, for example, make maximum use of the analytical capabilities of the active component 4th Psychological Operations Group to support a wide range of DOD and national-level requirements. The use of new technologies to enhance military psyops capabilities without significantly increasing personnel requirements also must be vigorously pursued. Overall, the key will be to demonstrate the cost-effective nature of military psyops and their utility in both peace and war.

Another serious challenge for military psyops will be to ensure that the psychological dimension receives appropriate attention by the new administration. To accomplish this, OSD, OJCS, and USSOCOM must coordinate their efforts to have psychological operations considered during the National Security Council's (NSC) national security strategy review. It is vital to get this foot in the door early to ensure that appropriate national-level emphasis is provided to DOD for continued revitalization of military psyops.

Similarly, DOD must continue its efforts to ensure an integrated interagency effort in this dimension. Building upon the solid achievements of DOD's military psyops' bilateral support of the United States Information Agency will help in this regard, but sorely needed is a senior-level NSC mechanism that can integrate and direct the information and psychological operations activities of the various agencies in a coherent manner. Only in this way will the United States be able to compete more effectively with the Soviet Union in the "war of ideas."

Finally, perhaps the biggest challenge facing military psyops advocates in the near future will be educating the new players in DOD. After the new secretary and his key advisers are all in place, this process will begin. It is crucial to do so because a number of important directives affecting military psyops must be revalidated or modified accordingly. Again, this is an area where effective coordination among OSD, the OJCS, and USSOCOM is critical if "top-down" emphasis in DOD is to be restored.

A key role in this endeavor must be played by the commander-in-chief, United States Special Operations Command. After having prevailed in the lengthy debate over whether psyops and civil affairs forces should be assigned to USSOCOM, he must now assume a major leadership role in implementing a revalidated DOD Psychological Operations Master Plan. He must ensure that psyops become a full partner in the special operations community, and that this capability receives its appropriate share of attention and resources. With regard to the latter, he, along with the assistant secretary of Defense for Special Operations and Low-Intensity Conflict, have important voices in the allocation of resources as members of the Defense Resources Board.

At the same time, it is important for the USSOCOM commander to recognize that military psyops continue to have broader requirements beyond the support of special operations forces and that psyops have unique personnel and training requirements. In this regard, the creation of a separate principal staff directorate for psyops and civil affairs in USSOCOM by the current commander, General James Lindsay (United States Army), is an encouraging sign. His leadership, and that of his successors, will be crucial to ensuring that this delicate balance is maintained, and that the revitalization of military psyops goes forward.

In summary, U.S. interest in military psychological operations during this century has been episodic, basically rising and falling during and after major conflicts. The Reagan administration interrupted this pattern by providing the impetus for a revitalization of U.S. military psyops during peacetime. The international and domestic conditions that made this possible have changed significantly, however, and military psyops face considerable challenges in continuing to rebuild their capabilities. The commander-in-chief, USSOCOM, and the assistant secretary of Defense for Special Operations and Low-Intensity Conflict must play key leadership roles in seeing that this revitalization continues, and that military psyops' unique capabilities be employed broadly to include support of national level activities during peacetime. A senior-level NSC mechanism is needed to integrate interagency efforts in the psychological dimension to ensure a coherent response to Soviet initiatives. Military psychological operations can, and should, play an important supporting role in this endeavor.

4

Soviet Active Measures and Propaganda: "New Thinking" and Influence Activities in the Gorbachev Era

Herbert Romerstein

Active measures is a Soviet term meaning influence operations. It refers to the methodology used by the Soviet Union to cause a target to take actions beneficial to Soviet interests. Active measures have historically been primarily malevolent. They can be, under proper circumstances, benevolent. During the present period of reduced international tensions, active measures may prove to be a combination of both.

Soviet active measures appear in three modes, black, white, and gray. Black, or covert, active measures are conducted by the KGB. These include forgeries, agents of influence, and covert media placements of disinformation stories. White, or overt, active measures are carried out by entities directed by the Ideology Department of the Central Committee of the Communist party of the Soviet Union (formerly called the Propaganda Department). These are the propaganda instruments of the Soviet apparatus, such as Radio Moscow, foreign language magazines, books, and newspapers, which replay the themes used throughout the world in KGB disinformation. Gray, or semicovert, active measures are conducted by those entities directed by the International Department of the Central Committee of the Communist party of the Soviet Union (CPSU). These include: those foreign Communist parties that are still under Soviet influence; a network of international front organizations, each targeted against a different constituency, that is, youth, students, women, the peace movement, clergy, trade unions, journalists, and so on; and the international departments of Soviet mass organizations (sometimes called social organizations) that maintain foreign contacts.

On September 30, 1988, a new level of CPSU control was created for all areas of the Soviet apparatus. Six commissions were formed under the Central Committee, three of which have responsibility for, among other things, elements of the active measures apparatus. They meet at least once every three months to advise the Central Committee on developments in their area, and, with the permission of the Politburo, direct the work of the relevant departments.[1]

The Ideological Commission, headed by Vadim A. Medvedev, directs the Ideology Department. Medvedev had previous experience in this work. From 1970 until 1978, he served as deputy head of the Propaganda Department of the Central Committee of the CPSU.[2] The Legal Policy Commission, headed by former KGB chairman, Viktor M. Chebrikov, directs the work of a number of Soviet government organizations including the KGB. The International Policies Commission, headed by a close Gorbachev associate, Aleksandr N. Yakovlev, oversees the work of the International Department. Yakovlev has had considerable experience in the West, having studied at Columbia University as an exchange student in 1959 and served from 1973 to 1983 as Soviet Ambassador to Canada.[3] He is believed to be the intellectual father of glasnost, perestroika, and the "new political thinking."

The Soviet apparatus for active measures did not emerge "full blown from the brow of Jove." Each element—black, white, and gray—developed from entities created in the early days of the Bolshevik Revolution. The KGB evolved from the Cheka and OGPU of Lenin and Dzerzhinsky. The International Department is the direct descendent of the Communist International founded in 1919. The Ideology Department evolved from Lenin's propaganda apparatus.

The interrelationship between the elements of the active measures apparatus was described in 1983 by Konstantin Chernenko. Speaking to a meeting of the Central Committee of the CPSU, he said:

> Comrades, our entire system of ideological work should operate as a well-arranged orchestra in which every instrument has a distinctive voice and leads its theme, while harmony is achieved by skillful conducting. The main demands on party leadership of ideological work are constantly to check the tone of propaganda against our policy goals and people's interests, and to ensure that "word becomes deed," as Lenin put it. Propaganda is called upon to embrace every aspect of social life and every social group and region and to reach every individual.[4]

Active measures reflect the policies of the Soviet government and party. The present line of the party, exemplified by glasnost and new political thinking, was developed to solve the pressing internal problems of the Soviet Union. According to an article in *Kommunist* by two senior officials of the Soviet Institute of the United States and Canada, the present crisis is so serious that the Soviet Union is in danger of losing its position as a world power. They wrote:

> Today our country clearly needs the greatest possible change in the system of foreign policy priorities in the entire postwar period.
>
> The success or failure of perestroika is a question of whether or not we shall be able to preserve our socialist state, the first in the world, as a great power or else find ourselves on the margin of history, weakened furthermore, considering the adverse trends which are bound to increase should our economic and political reform stop midway.
>
> Currently we are given an almost last chance of retaining our position as a great power and reorganizing this position on a new basis, in accordance with new parameters, above all by relying on socioeconomic, legal, cultural and spiritual components.[5]

Taking advantage of the good image created by the new line, the Soviets have applied it in their international relations. For example, speaking to the Congress of the Soviet Trade Unions, in February 1987, Gorbachev pointed out, "Especially since the January meeting of the Central Committee, the words 'perestroika' (restructuring) and 'glasnost' (openness) are considered everywhere abroad to be synonymous with the words 'progress' and 'peace'."[6] Not that they are synonymous, but that people abroad so consider them.

Gorbachev has actively sought to reinforce the perceived link between internal reforms and a more benign foreign policy. Speaking to an international peace forum in Moscow in the same month, he said, "Our foreign policy is today to a greater extent than ever before determined by domestic policy."[7]

There is evidence, however that the Soviet Union has not changed its view of the world but only its tactics. Speaking before the Twenty-seventh Congress of the Communist party of the Soviet Union, Gorbachev said:

> The "enlightened" 20th century is going down in history as a time marked by such outgrowths of imperialism as the most devastating wars, an orgy of militarism and fascism, genocide, and the destitution of millions of people. Ignorance and obscurantism go hand in hand in the capitalist world with outstanding achievements of science and culture. That is the society we are compelled to be neighbors of, and we must look for ways of cooperation and mutual understanding. Such is the command of history.[8]

The new improved Soviet image is designed to facilitate the policy of "cooperation and mutual understanding" with the capitalist society that Gorbachev portrayed in such unflattering terms. This helps explain the sharp change in the Soviet political line from confrontation to cooperation. From the Soviet viewpoint, circumstance and the relationship of forces require the new benevolent international stance.

On February 13, 1987, Gorbachev held a meeting with the leaders of Soviet mass media and propaganda organs. One of those in attendance was Yegor Yakovlev, the editor-in-chief of *Moscow News*, the most glasnostian of all Soviet publications.[9] Gorbachev provided the historical and theoretical justification for the new Soviet political line. According to Yakovlev,

> Frankly speaking, it came as a surprise when Gorbachev spoke of the Brest Peace Treaty, the treaty which Lenin had described as a disgraceful, oppressive, immeasurably severe, obscene, foul, and humiliating peace. What is the analogy here? The Brest Treaty exemplified how short-lived interests could be sacrificed for a historic turn to secure vital interests.

Yakovlev, in a footnote, described the treaty,

> The treaty Soviet Russia signed in March 1918 with Germany, Austria-Hungary, Bulgaria, and Turkey provided for German annexation of Poland, the Baltic region, and Parts of Byelorussia and Transcaucasia, and an indemnity of 6,000 million marks. Soviet Russia had to accept these terms because of the difficult situation in which it had found itself. The treaty made it possible for Soviet Russia to pull out of the First World War and gave it a respite for the strengthening of Soviet government. The treaty was annulled by the Soviet government in November 1918.[10]

Whether the new political thinking will be long-lived, only time will tell. At the present time, however, it plays a major role in the policies of the Soviet active measures apparatus. From July 25 to July 27, 1988, an extraordinary conference took place at the Soviet Ministry of Foreign Affairs. Every element of the active measures apparatus participated, including the KGB. In a break with past secrecy, extensive excerpts from the speeches were published.[11] The main report was delivered by Eduard Shevardnadze, the Soviet foreign minister, who pointed to "an increasing shift in efforts to ensure security from the sphere of military-political solutions to the sphere of political cooperation" in Soviet policy.[12] A shift, in other words, from the threat of force to political influence operations.

One speaker at the conference was Albert I. Vlasov, then first deputy head of the Propaganda Department of the Central Committee, CPSU.[13] He complained about, "the losses which our propaganda suffered in the international arena in the past." He pointed out however, that,

> The restructuring drive has turned around not only foreign public opinion, but also old stereotypes about the Soviet Union. That the stereotype of the "Soviet threat" and the "enemy image" has been steadily on the wane since April 1985 is to the credit not only of staff members of foreign policy departments and foreign policy propagandists. It is the direct result of *perestroika* and revolutionary changes in Soviet domestic and foreign affairs. It is these changes that are rapidly winning this country sympathies abroad.[14]

At the time of the conference, Vladimir A. Kryuchkov was the deputy chairman and head of the First Chief Directorate of KGB. He has since been made chairman of KGB. Speaking at the conference, Kryuchkov complained that, "In the past we frequently viewed the world in simplistic, general terms. . . . We poorly study and know people; we pay too little attention to the development of contacts with foreign political and public figures, and we are too sluggish and inconsistent in fighting for their hearts and minds. It is evidently easier to create enemies than to win supporters."

Kryuchkov said that, "Without an objective vision of the world, a vision of it without embellishments, clichés and stereotypes, all talk about the effectiveness of our foreign policy actions is mere phrase-mongering." He went on to say, "The Soviet security service is waging a stubborn and complicated struggle. We attach great importance to the human factor, to teaching staff members to think in new political categories."[15] These statements reflect the two roles of the KGB's first chief directorate. One is collecting intelligence against the external adversary and analyzing it objectively, "without embellishments, clichés and stereotypes." The other is covert active measures, the "stubborn and complicated struggle," which requires that the KGB officers "think in new political categories" to enable them to promote the line of the party.

Karen Brutents was deputy head of the International Department, Central Committee, CPSU, at the time of the conference. He has since been promoted to first deputy chief, responsible for foreign communist parties, international Soviet fronts, Soviet mass or social organizations which have contacts abroad, and relations with noncommunist parties and organizations. He boasted at the conference that not only have the Soviets improved their relations with left-wing foreign movements but also with more conservative organizations in the West. Referring to the Soviet mass organizations with foreign contacts, he said, "Formerly these organizations were mere commentators publicizing our policy. While retaining and, indeed, extending this function, they now have a role to play in shaping our international policy."[16]

Yevgeni Velikhov, vice president of the USSR Academy of Sciences, maintains extensive contacts with scientists in the West. He described to the conference how he and other Soviet scientists participated in the active measures campaign against the Strategic Defense Initiative (SDI). Even before President Reagan announced SDI, the Soviet apparatus was preparing arguments against the plan to defend the United States against Soviet nuclear missiles. According to Velikhov,

> In the middle of April we held a discussion of the Strategic Defense Initiative (SDI), as it was later called, at the USSR Academy of Sciences with all the members of the Academy. The discussion was not easy—we could not express all the arguments under the restrictions of secrecy, and the official position was not yet formulated, but the resulting document was good, and I would subscribe to it even today. It was used by the opponents of SDI among U.S. scientists during a debate in the U.S. Congress.
>
> We were prepared to look for meeting the U.S. challenge in a nontraditional way. The country's leadership, using to some extent the recommendations provided by Soviet scientists and the opinion of the world scientific community, formulated the concept of an asymmetrical effective response to possible U.S. actions. [This refers to the Soviet threat to develop nuclear weapons that could penetrate the SDI shield—HR.] That was a fairly sobering decision and, I am sure, it influenced U.S. public opinion and the Congress decisions on limiting the SDI budget and leaving the 1972 ABM Treaty unchanged.[17]

Valentina Tereshkova, the famous Soviet cosmonaut, is chairwoman of the Union of the Soviet Societies for Friendship and Cultural Relations with Foreign Countries. This is the umbrella group that unites Soviet friendship societies, each of which is targeted on a specific country, such as the USSR/USA Friendship Society, USSR/France Friendship Society, and so on. Tereshkova pointed out that,

> The very fact that spokesmen for public organizations were invited to the Foreign Ministry scientific and practical conference is evidence of a new attitude of our professional diplomacy to people-to-people democracy. Mikhail Gorbachev's report to the 19th All-Union CPSU Conference pointed out the increased role of the people in world affairs. It stressed that objective world processes should become a subject of constant scientific and public

> discussions involving the general public and its organizations. People-to-people diplomacy is a reality in today's world. Policy can no longer be shaped without regard to the activity of mass anti-war, anti-nuclear, environmental, youth, women's, friendship movements in all countries.
>
> We are directing all our efforts towards applying new political thinking, which expects nations to know the truth about each other, discard mutual prejudice, renounce the enemy image and extend informal ties. Growing links are characterized by a rediscovery of the Soviet Union on the part of the foreign public and, furthermore, by the possibility for Soviet people to know and understand the people of other countries better, to benefit from ideas generated by other cultures and intellectual traditions.
>
> In every sphere of our work, we are cooperating actively with the Soviet Foreign Ministry and the foreign ministries of the union republics.

Tereshkova explained the importance of influencing noncommunists in the West. Previously the friendship societies maintained contact primarily with foreign communists and communist sympathizers. She said,

> In helping new political thinking take root in practice, we seek opportunities for dialogue on a broader political basis than before and strive to promote contacts and cooperation with all who stand for preserving human civilization, for a nuclear-free and non-violent world. The fact that a long time we were confined within a narrow milieu, that is, mainly within the environment of like-minded people abroad, could not but affect the content and style of the work of the union and, indeed, its foreign counterparts. This robbed us to a considerable degree of flexibility, which is so necessary for expanding contacts with movements and organizations of a different ideological orientation.[18]

Genrikh Borovik is chairman of the Soviet Peace Committee. He is the brother-in-law of KGB Chief Vladimir Kryuchkov.[19] In Borovik's report to the Ministry of Foreign Affairs Conference, he complained,

> We have many difficult problems to contend with. A bureaucratic image of the Soviet Peace Committee has been created over many years. The Peace Committee itself, with its inflexibility and lack of resourcefulness, is a fault, too. And, of course, the image created by Western

> propaganda. Today a great deal is changing. Our multifaceted ties have been augmented tenfold. This is no exaggeration. Relations are developing with a very broad range of organizations, and not only with our friends, whom we are accustomed to dealing with. The motto "Those who are not with us are against us" is not suited for a dialogue on how to save the world by joint effort. We want the World Peace Council, too, to become a more pluralistic organization.[20]

The World Peace Council is an international Soviet front, directed by the Soviet Peace Committee on behalf of the International Department of the CPSU. Recent developments in the World Peace Council will be related later in this chapter.

New political thinking has not been universally accepted in the world communist movement. Not only has it been resisted in Eastern Europe and Cuba, but a number of foreign communist parties, particularly those noted for subservience to the CPSU, have also objected.

World Marxist Review—Problems of Peace and Socialism is the international communist movement's theoretical organ. Headquartered in Prague, it is directed by the International Department of the CPSU. Meetings sponsored by the magazine are designed to provide direction to the international communist movement without openly revealing this fact. In November 1981, two representatives of the Worker's party of Jamaica (the Communist party of that country) attended a meeting in Prague sponsored by *World Marxist Review.* Top level officials of all of the Communist parties attended this meeting. The Jamaican communists in a secret report to their party revealed,

> Why was the Conference at such a high level? To understand this, we need to understand that there is presently no formal center of the International Communist Movement. Due to the opposition of the Eurocommunists, the Chinese CP and their supporters in the movement (e.g. Romania), it is difficult to hold even meetings of C and WP (Communist and Worker's Parties). The last was held in 1969. The Journal and its regular Conferences, therefore, play an important role of providing for contact, exchange of views and formulation of international policy. Through their representatives in the Journal at Prague (very often CC [Central Committee] members) and in these conferences now being held every four years, important international questions can be tackled. This has to be done, unfortunately, without any formal acknowledgement of its role, which would intensify the charges that the WMR is acting as a "center" of the

movement.[21]

In December 1987, *World Marxist Review* held a conference at which four Soviet scholars from the Social Sciences Institute under the CPSU Central Committee[22] presented a paper on the new political line. There was considerable resistance to the positions taken by the Soviets. The Soviet paper contained the following formulations: "Joint action by today's progressive forces is based on universal human values central to the very existence of society. They have priority over the differences rooted in class and national interests." The paper suggested that armed force should not be used in class struggle because of "the concept of a non-violent world." The paper further said, "Today, revolutionary tasks should be tackled in the context of the struggle for survival, and this struggle concerns all people regardless of their class affiliation. . . . Paradoxically, the class adversary should become a partner in the efforts to solve this universal problem."

The dogmatists at the meeting responded in traditional fashion. Bert Ramelson, an old Stalinist hardliner from the British Communist party, complained:

> The only consistent thing about this paper is that its authors express ideas leading away from Marxism. The impression is that for the sake of new thinking, we should renounce the class struggle and that social change in the contemporary world should no longer take the form of revolution. As a Marxist, I cannot accept this interpretation. The struggle will go on, and that should be taken into account when we discuss new thinking. Class struggle cannot be turned on and off like a tap. . . . I cannot accept the idea of some supraclass values. Every class has its own values. . . . Partnership between imperialism and socialism is a concept from the right-wing vocabulary. That sort of lasting unity is unattainable. The right-wing Social Democrats may be satisfied, but the communist party will end up losing everything it has accomplished in its fight for the hegemony of the working class.

Hector Mujica of the Communist party of Venezuela was shocked by the suggestion that communists cease armed struggle. He said, "The USSR is doing all it can to preserve peace and achieve disarmament. But I cannot even contemplate the idea of the CPSU urging, say, the Farabundo Martí National Liberation Front to lay down their arms or the Chilean Communists to accept Pinochet's regime for the sake of universal peace and disarmament." Essop Pahad of the South African Communist party also complained about the concept of non-violent struggle and pointed out that "not only the socialist countries but also, say, the World Council of Churches, recognize our right to armed struggle." Egyptian Communist party representative Kamel Magdi was also outraged. He explained, "the use of force on our part is merely a legitimate response. Who can persuade the Arab peoples to give up their struggle against Israeli aggressors, to lay down their arms and surrender to the invaders?"[23]

Indeed, Foreign Minister Shevardnadze has indicated that the Soviets may very well reduce their subsidies to Third World countries, due to financial constraints. If significant foreign policy initiatives must be curtailed for financial reasons, it is very possible that Soviet support for insurgent, terrorist, and other violence-prone groups will also have to be curtailed. Shevardnadze explained the problem in a speech on November 1, 1988, to the Communist party organization in the Ministry of Foreign Affairs. He said that there were a number of things they would like to do in foreign policy that they will be unable to accomplish because of the status of the Soviet economy and budget restrictions. He indicated that there was a potential problem with those developing states friendly to the Soviet Union who would be affected by the reductions. He suggested making the changes slowly so that there would be "a controlled transition of relations." While it would be preferable, he said, to increase the subsidies, in fact it was impossible to avoid reducing the current levels.

Shevardnadze also criticized how foreign policy had been made in the past, without any public input. He said, "For decades we have made decisions in foreign policy which the people invariably greeted 'with warm approval'. They approved policies that caused them to be victims, that brought the country to ruin, that created a caricature of the [Soviet] people themselves [in the eyes of the world]."[24]

The hardliners in the leadership of the foreign Communist parties have even complained about glasnost, the Soviet policy of telling the truth about some of their shortcomings. Gus Hall, the head of the Communist party in the United States, complained in a speech at a Communist party meeting in Cleveland, Ohio, on August 2, 1988, "I've been battling with the *Moscow News* now for a year or two, I met with some leading comrades there [in Moscow]. . . . I said why don't you remove the editor of the *Moscow News*, 'cause he's publishing a lot of nonsense, and I said if you can't do that, burn down the goddamn building, I mean get rid of it. But they laughed and they haven't done it. *Moscow News* is the worst of their publications."

During a *World Marxist Review* meeting in Prague on the subject of perestroika, the American Communist party representative, James West, echoed the complaint of his leader Gus Hall,

> Our party understands the need for perestroika and glasnost and desires their complete success. This would clearly establish the superiority of socialism over capitalism and have a revolutionizing effect. There is, however, a

> serious problem with some articles which have appeared in the *Moscow News*, *New Times*, and *World Marxist Review*. I think that a special responsibility rests on Soviet comrades for the statements which are made in the Soviet and communist press and which are often regarded in the United States as reflecting the official position of the Soviet Union. Due account should be taken of the response to these statements abroad, on the part of both Communists and non-Communists.

West was answered by Yur Krasin, the rector of the Social Sciences Institute of the Central Committee of the CPSU,

> If I understood him [West] correctly, he objects to the fact that some of our newspapers and periodicals sold abroad create more problems for the U.S. Communists than they do anything useful. He said the Soviet press bore a special responsibility. I'll begin by answering this quite forthrightly and then explain my answer. I am convinced that there cannot be one kind of truth for the Soviet people and another for the public, for the Communists abroad. Truth is one, it is indivisible, and it cannot be measured out differently to different audiences.

Krasin also responded to the complaint of Jose Riva of the Dominican Communist party, who worried that, "the efforts to tackle the problems of peace and disarmament may somehow push into background solidarity with the peoples striving for national liberation." Krasin said, "I am far from trying to deny the role of the liberation movements themselves, but the problem assumes an international dimension in which it is intertwined with the struggle for a non-violent world, for disarmament. If you are committed to revolution, you have to think in terms of world, as well as domestic politics nowadays."[25]

The Soviet new political thinking plays a major role in its current relations with the United States. In addition to the INF Treaty, which for the first time actually reduced nuclear weapons rather than establishing limits on the creation of new ones, the Soviets have taken a number of positive steps. We have seen this in the area of free exchange of information, where the Soviets have not only stopped jamming Voice of America, but also Radio Liberty and Radio Free Europe. The purely negative image of the United States in the Soviet press has been replaced by a more balanced tone. Albert Vlasov, now Chairman of *Novosti*, told the July conference at the Soviet Foreign Ministry,

> No one is calling upon the Soviet media, as was the case in the past, to portray Western reality and the foreign

> policies of the USA and its allies in exclusively black tones, but if we are speaking candidly about what is bad in our country, we must show just as openly the many negative phenomena there, too. This does not mean that we should conceal from our public the good elements there. We have seen for ourselves that such propaganda is not believed, and given the diverse means of obtaining information today, both Soviet people and the foreign public will simply choose alternative sources of information, and not ours.

Vlasov also said, "At the initiative of the Soviet side, complicated dialogue has been initiated with the American side; a search is underway for criteria which is making it possible to draw a dichotomy between international information and misinformation, and criteria for certain norms of mutual relations in the information sphere. This approach is in accord with the spirit of peaceful coexistence and confidence-building."[26]

Vlasov was referring to an ongoing series of discussions between U.S. and Soviet officials organized by Charles Z. Wick, then director of the United States Information Agency, and Valentin M. Falin, then chairman of Novosti Press Agency. (Falin is now head of the International Department and Vlasov now heads *Novosti*.) In September 1988, Vlasov was an active participant in a meeting in Moscow where, at the suggestion of Falin, an "early warning system" was established to help stop deliberate disinformation.

A major role at the September Moscow meeting was played by Vadim P. Perfilyev, the deputy chief of the Information Department of the USSR Ministry of Foreign Affairs. Speaking on November 21, 1988, at the U.N., Perfilyev explained Soviet concerns in these words,

> Sometimes we witness a pointed, and I would say, a painful response of a country to the information that was disclosed about it abroad. Not infrequently what follows are categorical public recriminations and allegations about a deliberate dissipation of misinformation, about unfairness and etc. The responding sides have nothing against using the phraseology of "cold war." Maybe, it is time that here and now we dropped confrontational reflexes and looked for new ones, that would differ from cheap propaganda exchanges between specialists in the field of information.[27]

The Soviets are clearly concerned that the exposure of Soviet disinformation stories by the United States is damaging, however slightly, the positive image that the Soviets have created through glasnost and "new political thinking." The U.S. position has consistently been that the end of Soviet disinformation campaigns would result in the end of the U.S. exposure of these campaigns.[28] *The New York Times* of October 25, 1987, reported that Gorbachev complained to Secretary of State George Schultz about the 1987 State Department report. Schultz responded by pointing to Soviet disinformation stories about the United States in explaining why the United States had to publish the truth about these issues.

A reduction in the level of disinformation appearing in the Soviet press, that is, part of the white active measures apparatus, has clearly taken place. However, it has not decreased in black active measures conducted by KGB and in gray active measures directed by the International Department of the CPSU. The most recent example was the bizarre story that the United States imports Latin American babies for the purpose of dismembering them so that their spare parts can be used for organ transplants. This false story continues to be promoted by an international Soviet front, the International Association of Democratic Lawyers. The agitation of this front aided by a representative of the French Communist party to the European Parliament resulted in that body voting to investigate the charges in September 1980. The vote was taken while most members were absent and has resulted in substantial embarrassment to the highly respectable European Parliament. The misuse of their name to circulate the story throughout the world was used to damage the prestige of the United States by the black and gray elements of the Soviet active measures apparatus.

In July 1988, another international Soviet front promoted a forged document supposedly from the National Security Council. This forgery claimed that at an April 8, 1986, meeting of the NSC, President Reagan ordered a campaign to "destabilize" Panama. In January 1989, the International Organization of Journalists, an international Soviet front, repeated the false charges referring to the forged National Security Council document.[29]

During an international human rights conference held in Geneva in March 1989, a document was distributed to the delegates which alleged that there were political prisoners in the United States. The list of so-called political prisoners consisted primarily of people convicted of violent crimes and terrorism. A number of these criminals came from well-known terrorist organizations, such as the Weather Underground and the Puerto Rican FALN, which was responsible for the bombing of a New York restaurant in which four people were killed. The list was distributed by a group called National Campaign for Amnesty & Human Rights for Political Prisoners. They were assisted in the distribution by the International Association of Democratic Lawyers, a Soviet front. The national campaign, also called the Movement Support Network, is a joint project of the Center for Constitutional Rights and the National Lawyers Guild. The latter is the U.S. affiliate of the International Association of Democratic Lawyers. A few weeks before the list was distributed in Geneva, the Indonesian daily newspaper *Merdeka* of

February 5, 1989, carried a story about alleged political prisoners in the United States based on information provided by the U.S. organizations. The article was provided by *Novosti.* A press release of the supposed human rights supporters in the United States makes it clear that those they are defending are actually terrorists. They said, "Political prisoners are those whose incarceration is the direct result of actions taken or expressions made to further a political or social vision."[30]

A close relationship has always existed between gray and black active measure elements. During the 1920s, the head of the foreign section of the Soviet Intelligence Service, the OGPU, was Moishe Abramovitch Trilisser. At the same time, he served as a high-level functionary of the Communist International, the gray active measures apparatus at its time. In 1924, he was a member of the Mandate Commission at the Fifth Congress of the Communist International, where he served as the Deputy to the veteran CPSU functionary Osip Piatnitsky, who headed the Organizational Bureau of the Comintern.[31] The next year, together with Piatnitsky, Trilisser again served as a member of the Mandate Commission at a meeting of the Enlarged Executive Committee of the Communist International.[32] When Piatnitsky fell victim to Stalin's purges, Trilisser replaced him as head of the Organizational Bureau. At the Seventh World Congress of the Comintern, he was elected to the Executive Committee, under the name M. A. Moskvin.[33] In 1971, an official Soviet history of the Comintern identified Moskvin as Trilisser.[34]

In a similar relationship, Vladimir Kryuchkov, the present head of KGB, serves on the International Policy Commission of the Central Committee, CPSU, the guiding force of today's gray active measures apparatus. Kryuchkov spent many years as head of the First Chief Directorate of KGB, which carries out intelligence operations abroad. He held essentially the same position in the intelligence service as that held by Trilisser in the 1920s.

The 17 international Soviet fronts form a major element of the gray Soviet active measures apparatus.[35] They are directed by the International Department of the Central Committee of the CPSU. They meet at least once a year to receive direction and themes and activities. In addition each front has a representative working on behalf of the International Department stationed at its headquarters to provide day-to-day guidance.

Most of the international Soviet fronts were established shortly after World War II. A few are of more recent vintage. The current fronts are the direct descendants of a network of Soviet fronts that operated in the 1920s and 1930s. Then, as now, the fronts pretended that they represented large numbers of people throughout the world. One of the prewar fronts, the Red International of Labor Unions, (RILU) claimed to represent millions of trade unionists. It was the ancestor of the present front, the World Federation of Trade Unions (WFTU). In 1985, the WFTU published its official history. It revealed that, "Between 1935 and 1936, the RILU ceased to exist."[36] Apparently no one, not even its millions of members noticed that the organization had disappeared. Its successor does not even know exactly when this disappearance took place.

The World Peace Council is the best known of the international Soviet fronts. It is directed by the International Department through the Soviet WPC affiliate, the Soviet Peace Committee. For many years the head of the Soviet Peace Committee and guide of the WPC was Yuri Zhukov. During his tenure, he succeeded in antagonizing most of the West European noncommunist peace activists. On December 2, 1982, Zhukov addressed a letter to the peace movements of Western Europe. He insisted that those who advocated eliminating both Eastern and Western nuclear weapons were not legitimate peace activists, since only Western and not Soviet nuclear weapons were a threat to peace. He refused to allow the Soviet Peace Committee to collaborate with such backsliders.[37] As a result, the SPC and the WPC were left talking to themselves. Only communists or inveterate fellow travelers would work with them.

The Trotskyites often accurately characterize their Stalinist cousins. In 1951, James P. Cannon, the head of the U.S. Trotskyites, described a meeting of the World Peace Council as "the recent Warsaw Peace Conference of professional fellow travelers, congenital stooges and moon-struck clergymen steered, like all such gatherings, by hard-faced jockeys from the Stalinist riding stables."[38]

Romesh Chandra, a longtime member of the Central Committee of the Communist party of India, has served for decades as president of the World Peace Council. As a hardline Soviet apologist, his reputation among West European peace activists is no better than that of Yuri Zhukov. In 1986, the WPC and the other fronts began an attempt to change their poor images. At the April 1986 session of the World Peace Council, in Sofia, Bulgaria, it became clear that new leadership was needed. Chandra spoke up and admitted his deficiencies. He said of himself, "The criticisms made of the President's work require to be heeded and necessary corrections made." He mentioned the proposal to create the office of secretary-general, who would run the day-to-day work of the operation. Chandra suggested that the changes not be made immediately but that a committee be set up to discuss them and report back some eight months later.[39]

Yuri Zhukov then stood up to speak. He also mentioned the proposal to create a new post of general-secretary, and said, "We, in the Soviet peace movement, have decided upon a comprehensive discussion to support this proposal. We hope it will be met with approval at the session." Indeed, all of the Soviet suggestions were "met with approval" including one that the new post, which would take much of the power away from Chandra, would be held by a West European, "preferably from the host country of the WPC headquarters." As Helsinki is the headquarters of the WPC, the post was given to

Johannes Pakaslahti, an activist in the hardline, pro-Stalinist faction of the Finnish Communist party. Zhukov suggested that the "changes should be made and adopted here."[40] They were, of course.

Tair Tairov formerly served as the Soviet representative on the staff of the World Peace Council in Helsinki. In an article in *Komsomolskaya Pravda*, on June 25, 1988, Tairov complained about the leadership of the Soviet Peace Committee. He pointed out that in the beginning of 1987 the former chairman of the Soviet Peace Committee [Yuri Zhukov] retired. Tarov referred to this as, "a commendable fact." He was replaced by Genrikh Borovik. But Tairov complained that Borovik was elected at a meeting where he was the only candidate and received a unanimous vote.

Tairov is a reformer who has opposed the entrenched leadership of the Soviet Peace Committee and World Peace Council. He has taken advantage of the difficulties that the WPC has had in improving its image among noncommunist peace activists. In an interview that appeared in the organ of the Finnish Peace Committee, *Rauhan Puolesta* for September 1988, Tairov suggested that new and younger leaders are needed in the World Peace Council. As an example of the stagnation in the WPC, Tairov cited the fact that the WPC gave medals to Brezhnev and Brezhnev gave medals to the WPC.

The problems of the WPC were not solved by bringing a new Stalinist, Pakaslahti, to share power with the old Stalinist, Chandra. In fact, the two have fought for the past three years. On January 12, 1988, the Finnish Communist party newspaper *Kansan Uutiset* reported that Chandra had sent a letter to the affiliated organization of the WPC announcing that Pakaslahti is no longer the secretary-general of WPC. Pakashlahti, in turn, claimed that he still holds the post and would remain in it until the next WPC meeting held in Athens in December 1989.

Despite the difficulties of the international Soviet fronts, they have become even more important to the Soviet active measures apparatus. As was pointed out in the case of the baby parts disinformation story, they are needed to spread disinformation when the official Soviet media reduces its level. They are also needed in other forms of influence operations.

The fronts, calling themselves "closely cooperation NGOs" (non-governmental organizations), are active at the United Nations where they participate in an organization called the Conference of Governmental Organizations (CONGO) in consultative status with the United Nations Economic and Social Council. Prior to September 1988, the fronts held five seats on CONGO's 18-member board. A sixth organization collaborated with them.[41]

The new CONGO board has 20 members, elected in September 1988. It contains seven Soviet fronts, or "closely cooperating NGOs," and one collaborator.[42] CONGO and the fronts have been pressing the United Nations to provide a greater role for the NGOs in U.N.

activities. To use an old Soviet phrase, "it is no accident" that in his December 7, 1988, speech to the U.N. Gorbachev said, "We in Moscow are pleased that more and more government, political, party and public leaders, and—I would like to particularly emphasize this—scientists, cultural figures, representatives of mass movements and various churches, activists of what is called people's diplomacy, are prepared to shoulder the burden of general responsibility." He went on to suggest, "In this context, I think the idea of convening an Assembly of Public Organization on a regular basis under the aegis of the United Nations also deserves consideration."[43]

The International Department of the Central Committee of the CPSU directs the international Soviet fronts as well as the activities of the Soviet Friendship Societies. They also direct the Rodina Society, which has responsibility for contacts with Soviet citizens abroad. A recent article in *Literaturnaya Gazeta* by Leonid Pochivalov, a vice-president of the Rodina Society, stretched a hand out to those émigrés who had been previously considered traitors. He wrote that, or course, his society maintains contact with those émigrés whom he described as sympathizers" and "progressive émigrés." He complained that they had frightened away many honest people who could be won over to support Soviet positions.[44]

The Soviets seldom admit that the International Department directs these organizations. However, they occasionally slip up, as when recently Valentin M. Falin, head of the International Department, was nominated to the Supreme Soviet as a representative of the Soviet Friendship Societies and the Rodina Society.[45]

The covert Soviet active measures apparatus continues to engage in disinformation. Newspapers such as *Blitz* and *Patriot* in India and *Ethos* in Greece, which have been well known for years as purveyors of Soviet disinformation continue to spread it. Dr. Jacob Segal of East Berlin, who was used by the Soviets to spread the false story that AIDS was deliberately created in U.S. laboratories, continues to spread that disinformation despite the fact that it no longer is carried in the Soviet media.[46]

Although most of the Soviet media have reduced the flow of disinformation, one medium, pamphlets and booklets, continues to spread both disinformation stories and enemy image propaganda against the United States. A 1987 pamphlet on the U.S. Constitution presented the following image of the United States:

> The anti-popular nature of imperialism reveals itself in the unprecedented efforts made in the United States to destroy human beings morally, to kill everything human in people at home and abroad. A robot programmed for violence and destruction; a robot without a heart, charged with hatred for all and incapable of entering into any normal human relations; a kind of addict without an intellect,

> unable to rise in his motives above the most primitive pornography—such is the typical "ideal" of an individual as advertised by contemporary mass culture in America, especially those schools and trends of mass culture that enjoy particular support from the present administration [Reagan administration].[47]

A 1988 Soviet booklet, *The CIA in Asia*, contained a variety of false stories of alleged CIA activities in India and Afghanistan. The most bizarre was one quoting from the Indian newspaper *Blitz*, a frequent purveyor of KGB disinformation stories.

> A *casus belli* was provided on February 14, 1979, when the U.S. Ambassador to Afghanistan, Adolf Dubs, was assassinated by terrorists. "It is being said," wrote *Blitz*, "that Ambassador Dubs was 'sacrificed' at the insistence of the CIA in order to bring about a total rift between Washington and Kabul by pinning the blame on the Tarakki regime." That opinion was shared by many in Asian democratic circles. There are certain occurrences that show Dubs had been sacrificed to a monstrous CIA plan.[48]

In fact, the death of Ambassador Dubs took place when he was taken hostage by terrorists in Kabul. An Afghan police unit confronting the terrorists opened fire when ordered to do so by their Soviet advisers. U.S. diplomats at the scene had pleaded with the Soviets to try to "talk down" the terrorists rather than risk the life of the ambassador in a firefight. The ambassador died when the American advice was ignored by the Soviet advisors.

One recent Soviet booklet is a virtual smorgasbord of disinformation stories. Titled *Once Again About the CIA*, this publication blames the U.S. intelligence agency for the deaths of a variety of foreign leaders, some of whom died violently and others of natural causes. The most outrageous story in this booklet blames the CIA for the 1978 murder of Congressman Leo Ryan and over 900 members of the Jonestown cult, who committed suicide or were killed on the orders of their insane, procommunist leader, James Jones. According to the booklet, "The CIA and other U.S. special services committed two massacres on November 18th: they killed Congressman Leo Ryan and several newsmen accompanying him and then slaughtered all the members of the Jonestown commune."[49] The Soviet Embassy in London advertised the sale of this booklet in January 1989.[50] The same false charges about Jonestown were made in a Soviet book published in 1987 in both Russian and English, titled *The Jonestown Carnage: A CIA Crime*.[51]

It is hard to imagine that anyone could believe so ridiculous a story. Ten years ago, television news throughout the world broadcast pictures of the Jonestown massacre and audio tape recordings of Jim Jones ordering his followers to kill one another and commit suicide. The Soviets apparently believe that some audiences in the world are so anti-American that they would believe even this false story. A Soviet textbook used in training foreign communist propagandists explained this concept,

> Knowing the audience, one can consciously "portion out" the rational and emotional elements in propaganda activity. It is of course a difficult task, determining the correlation between these elements in each concrete instance. A great deal depends here on the experiences and professional skill of the propagandist. Success will be determined here first and foremost by the knowledge of the socio-psychological features of the audience in all its parameters and aspects.[52]

The textbook also explained the use of *ad hominem* arguments.

> Situations sometimes arise in practical propaganda in which the logical mode (arguments *ad rem*) do not produce the necessary effect, despite the convincingness of the arguments and correctness of the propagandist's propositions. The psychological mode (arguments *ad hominem*) prove to be more effective, for the propagandist takes into account the usual course of reasoning and conclusions to which the listener resorts proceeding from his interests and convictions. Psychological arguments make the propagandist's words more convincing, comprehensible, and clear.[53]

The textbook was obtained in Moscow in 1985. Two years earlier, a document was found in Grenada which contained a report from Grenadan students at the Higher Party School of the CPSU. This school, also called the Lenin School, has existed in Moscow since the 1920s to train foreign communists. The report contained a listing of the courses taken by students, including "Social Psychology and Propaganda."[54]

U.S. officials complained to the Soviets about the crude disinformation contained in the pamphlet, *Once Again About the CIA*. A spokesman for the U.S. Embassy in Moscow, Richard Gilbert, told Associated Press, "We have discussed this book with the Soviets as a particularly egregious example of what we call disinformation. . . . It contains outright falsehoods, and we've told them we don't think the continued distribution of this book is in keeping with the increasing positive tone of our relationship."

When AP contacted Yevgeny A. Posdnyakov, deputy editor-in-chief of Novosti's North American Department, he defended the booklet and claimed it was part of the campaign for greater frankness and wider debate launched by Gorbachev. Posdnyakov said, "There is a need to write about everything so people will make up their own mind and understand things. . . . The demand that some zones be free from inquiry by journalists would be an attack on glasnost.[55]

Posdnyakov has been an active participant in the bilateral U.S.-Soviet discussions on the problem of disinformation. Another active participant, Leonid Dobrokhotov, of the Ideology Department, made a similar statement in an article in *New Times*. He claimed that the Soviet government and the Communist Party of the Soviet Union, Central Committee, are not responsible for what their official media publish, "in the period of democratization and glasnost."[56]

These views of glasnost seem to contradict the statements made by Gorbachev. He did not say that glasnost was a license to lie. What Gorbachev told a July 1987 meeting of Soviet media people was, "I am asking you to rise above emotions and convenient stereotypes. . . . We tell the truth and nothing but the truth."[57]

While crude disinformation continues to appear in the black and gray modes, and occasionally in the white, most Soviet white propaganda conforms to the dictates of the "new political thinking." But even the most friendly Soviet propaganda slogans have a purpose that may not be immediately apparent. For example, a current Soviet theme for U.S. audiences is that the two countries have developed a special relationship, indeed that the United States is the main partner of the Soviet Union in eliminating the threat of nuclear war. The concept of main partner, of course, excludes Western Europe. The primary slogan for Western Europe is the "Common European Home." This, of course, excludes the Americans. A combination of these two seemingly benevolent themes is aimed at achieving the long-term Soviet goal of splitting the Atlantic Alliance.

So important is the "Common European Home" theme that it was the major subject of discussion on January 6, 1988, when a delegation of the Communist Group in the European Parliament met with the then head of the International Department, Anatoly Dobrynin. The delegation included communists from Italy, Greece, Denmark, Spain, Portugal, and France. According to an official Soviet report of the meeting, "Particular attention was paid to the idea of the 'European home' advanced by Mikhail Gorbachev. Despite the fact that it concerns states belonging to different social systems and opposite military-political groupings, as well as neutral and nonaligned states, this concept reflects the integrity of Europe, the historical and cultural community of its peoples, the interdependence of the problems waiting for their resolution, and the need to create a stable and comprehensive system of security and peaceful cooperation."[58]

The Ideology Department of the Central Committee of the CPSU has the responsibility for both directing the propaganda organs of the

Soviet Union and of providing the party line and themes for all of the active measures elements. The head of that department is Aleksandr Kapto, who was appointed in October 1988. He has worked in both the party and government apparatus, holding the post of secretary of the Central Committee of the Communist party of the Ukraine and from 1986 to 1988 was Soviet ambassador to Cuba.[59]

Kapto made a report to the January 1989 meeting of the Ideology Commission on behalf of the Ideology Department. An official Soviet account on the commission session revealed,

> The Commission also listened to an information report delivered by A. S. Kapto on the functions and structure of the CPSU Central Committee Ideological Department. It was remarked that, though founded on the basis of three departments that used to exist—those of Propaganda, Science and Educational Institutions, and Culture—now that these have been virtually halved in size, the Central Committee Ideological Department is basically a new subdivision within the apparatus of the Party Central Committee which is intended effectively to carry out the CPSU line in the realm of ideology, to promote fulfillment of the Central Committee Politburo decisions, and to assist in the Commission's work on ideology.[60]

In an article in *Pravda*, February 20, 1989, Kapto described the role and functioning of his department. He said,

> Some Central Committee commissions, including the Ideology Commission, were created in accordance with the 19th All-Union Party Conference's decisions. The department works under the Ideology Commission's direct leadership. . . .
>
> The structural elements of the department have been amalgamated on the basis of six subdepartments, each of which has a clearly defined functional purpose. . . .
>
> The subdepartment of fundamental scientific research includes the sectors of the comprehensive problems of the development of science, philosophical, economic, and historical sciences and party scientific and educational institutions. . . .
>
> The party propaganda subdepartment includes the sectors of the theory and practice of ideological work and party study as well as groups of response organizers and lecturers. . . .
>
> The mass media subdepartment . . . includes the sectors of the periodical press, television and radio broadcasting, book publishing, and mass media

> effectiveness. Its main task is the implementation of the CPSU Central Committee guidelines on the leadership of the mass media and on increasing the efficiency of their work. . . .
>
> The subdepartment on foreign political information and international ties has sectors of analysis and long-term programs, current information, humanitarian problems, and international ties . . . the needs of the restructuring of the foreign policy sphere and the advancement of the priority of general human interests and values sharply redirect our informational and ideological activity in the international arena and place all aspects of the new political thinking at the center of this work. . . .
>
> The newly created subdepartment of the problems of the training and education of young people consists of the sectors of general secondary education, vocational training, Marxist-Leninist education, and the comprehensive problems of the upbringing of young people. . . .
>
> The culture and arts subdepartment consists of the sectors of the comprehensive problems of the development of culture, literature, arts, mass cultural work, and links with the public.[61]

This highly structured apparatus provides the themes and slogans for every element of the Soviet propaganda apparatus, at home and abroad. The active measures carried out pursuant to the guidance of the Ideology Department can be benevolent, malevolent, or a combination of both. Crude disinformation and enemy image propaganda appear side-by-side with friendly gestures and gentle concepts. The Ideology Department directs them all. This provides the Soviet Union with an efficient and effective machine to influence a variety of target audiences.

COMMISSIONS OF THE CENTRAL COMMITTEE, CPSU, RESPONSIBLE FOR ACTIVE MEASURES

Ideology Commission

Vadim A. Medvedev, chairman of the Commission.

Members of the Commission:

A. N. Aksenov, member of the CPSU Central Committee and chairman of the USSR State Committee for Television and Radio Broadcasting.
V. G. Afanasyev, member of the CPSU Central Committee and chief editor of *Pravada*.
L. K. Balykova, member of the CPSU Central Auditing Commission and a secondary school teacher from Omsk oblast.
A. A. Belyayev, member of the CPSU Central Auditing Commission and chief editor of *Sovetskaya Kultura*.
Ya. Ya. Vagris, member of the CPSU Central Committee and first secretary of the Latvian Communist Party Central Committee.
N. V. Gellert, member of the CPSU Central Committee and secretary of the Ukrainian Communist Party Central Committee.
A. T. Gonchar, candidate member of the CPSU Central Committee and sovkhoz machine operator from Tselinograd oblast, Kazakh SSR.
Yu. N. Yelchenko, member of the CPSU Central Committee and secretary of the Ukrainian Communist Party Central Committee.
V. G. Zakharov, member of the CPSU Central Committee and USSR minister of Culture.
T. G. Ivanova, candidate member of the CPSU Central Committee and deputy chairman of the RDFDR Supreme Soviet Presidium.
A. S. Kapto, member of the CPSU Central Committee and head of the CPSU Central Committee Ideology Department.
Yu. S. Karabasov, candidate member of the CPSU Central Committee and secretary of Moscow CPSU Gorkom.
V. V. Karpov, member of the CPSU Central Committee and first secretary of the Board of the USSR Union of Writers.
V. A. Kuptsov, member of the CPSU Central Committee and first secretary of the Vologda CPSU Obkom.
G. I. Marchuk, member of the CPSU Central Committee and president of the USSR Academy of Sciences.
M. F. Nenashev, candidate member of the CPSU Central Committee and chairman of the USSR State Committee for Publishing Houses, Printing Plants, and the Book Trade.
D. I. Patiashvili, member of the CPSU Central Committee and first secretary of the Georgia Communist Party Central Committee.

G. L. Smirnov, candidate member of the CPSU Central Committee and director of the Institute of Marxism-Leninism of the CPSU Central Committee.
S. P. Tkachev, member of the CPSU Central Auditing Commission and chairman of the Board of the RSFSR Union of Theater Workers.
M. A. Ulyanov, member of the CPSU Central Auditing Commission and chairman of the Board of the RSFSR Union of Theater Workers.
I. T. Frolov, member of the CPSU Central Committee and aide to the general secretary of the CPSU Central Committee.
T. N. Khrennikov, candidate member of the CPSU Central Committee and first secretary of the Board of the USSR Union of Composers.
G. A. Yagodin, member of the CPSU Central Committee and chairman of the USSR State Committee for Public Education.
R. G. Yanovskiy, member of the CPSU Central Auditing Commission and rector of the Academy of Social Sciences of the CPSU Central Committee.

International Policy Commission

Aleksandr N. Yakovlev, chairman of the Commission

Members of the Commission

G. A. Arbatov, member of the CPSU Central Committee and director of the Institute of the USA and Canada of the USSR Academy of Sciences.
S. G. Arutyunyan, first secretary of the Armenian Communist Party Central Committee.
S. F. Akhromeyev, member of the CPSU Central Committee, formerly chief of the General Staff of the USSR Armed Forces and first deputy USSR defense minister.
B. T. Batsanov, candidate member of the CPSU Central Committee and head of the Secretariat of the chairman of the USSR Council of Ministers.
A-R. Kh. Vezirov, first secretary of the Azerbaijan Communist Party Central Committee.
Ye. P. Velikhov, candidate member of the CPSU Central Committee and vice president of the USSR Academy of Sciences.
V. I. Valjas, first secretary of the Estonian Communist Party Central Committee.
V. M. Kamentsev, member of the CPSU Central Committee and deputy chairman of the USSR Council of Ministers.
A. G. Kovalev, member of the Central Auditing Commission of the CPSU and first deputy USSR foreign minister.

V. A. Kryuchkov, member of the CPSU Central Committee and chairman of the USSR Committee on State Security.

I. D. Laptev, candidate member of the CPSU Central Committee and chief editor of *Izvestiya*.

P. K. Luchinskiy, candidate member of the CPSU Central Committee and second secretary of the Tajik Communist Party Central Committee.

Yu. A. Manayenko, member of the CPSU Central Committee and first secretary of the Lipetsk CPSU Obkom.

V. M. Nikiforov, candidate member of the CPSU Central Committee and USSR deputy minister of Foreign Affairs.

R. N. Nishanov, first secretary of the Uzbekistan Communist Party Central Committee.

G. V. Novozhilov, member of the CPSU Central Committee and general designer at the S. V. Ilyushin Moscow engineering works.

Ye. M. Primakov, candidate member of the CPSU Central Committee and director of the Institute of World Economics and International Relations of the USSR Academy of Sciences.

G. I. Revenko, member of the CPSU Central Committee and first secretary of the Kiev Obkom of the Ukrainian Communist Party.

V. V. Tereshkova, member of the CPSU Central Committee and chairman of the Presidium of the Union of Soviet Societies for Friendship and Cultural Relations with Foreign Countries.

G. I. Ulanov, candidate member of the CPSU Central Committee and first secretary of Kaluga CPSU Obkom.

V. M. Falin, candidate member of the CPSU Central Committee and head of the international department of the CPSU Central Committee.

A. S. Chernyayev, member of the CPSU Central Committee and aide to the general secretary of the CPSU Central Committee.

Legal Policy Commission

Viktor M. Chebrikov, chairman of the Commission, former chairman of KGB

Members of the Commission:

V. V. Bakatin, member of the CPSU Central Committee and USSR minister of Internal Affairs (M.V.D.).

Yu. P. Batalin, member of the CPSU Central Committee and deputy chairman of the USSR Council of Ministers.

F. D. Bobkov, member of the CPSU Central Committee and first deputy chairman of the USSR KGB.

L. F. Bobykin, candidate member of the CPSU Central Committee and first secretary of Sverdlovsk obkom.

A-M. K. Brazauskas, first secretary of the Lithuania Communist Party Central Committee.

A. N. Girenko, candidate member of the CPSU Central Committee and first secretary of the Crimean obkom of the Ukraine Communist Party.

V. A. Zatvornitskiy, member of the CPSU Central Committee and leader of an integrated team of Moscow Main Committee for Construction.

B. V. Kravtsov, candidate member of the CPSU Central Committee and USSR minister of Justice.

A. D. Lizichev, member of the CPSU Central Committee, chief of the Main Political Directorate of the Soviet Army and Navy.

A. A. Logunov, member of the CPSU Central Committee and vice president of the USSR Academy of Sciences.

K. Makhkamov, member of the CPSU Central Committee and first secretary of the Tajikstan Communist Party Central Committee.

V. M. Mishin, member of the CPSU Central Committee and AUCCTU (Trade Union) secretary.

V. P. Mysnichenko, member of the CPSU Central Committee and first secretary of Kharkov Ukrainian Communist Party Obkom.

I. Kh. Romazan, candidate member of the CPSU Central Committee and director of the Magnitogorsk metallurgical combine named after V. I. Lenin.

N. F. Rubtsov, member of the CPSU Central Auditing Commission and chief of the Secretariat of the USSR Supreme Soviet Presidium.

V. I. Terebilov, member of the CPSU Central Committee and chairman of the USSR Supreme Court.

M. I. Khaldeyev, member of the CPSU Central Auditing Commission and chief editor of *Partiynaya Zhizn* magazine.

A. A. Khomyakov, member of the CPSU Central Committee and first secretary of the Saratov CPSU Obkom.

Ye. M. Chekhrain, candidate member of the CPSU Central Committee and deputy chairman of the RSFSR Council of Ministers.

Ye. A. Shulyak, member of the CPSU Central Committee and press-forge operator from the Minsk "60th Anniversary of Great October" motor vehicle works.

(List updated from *Pravda*, November 29, 1988—in Russian)

Soviet Organizational Structure for Active Measures

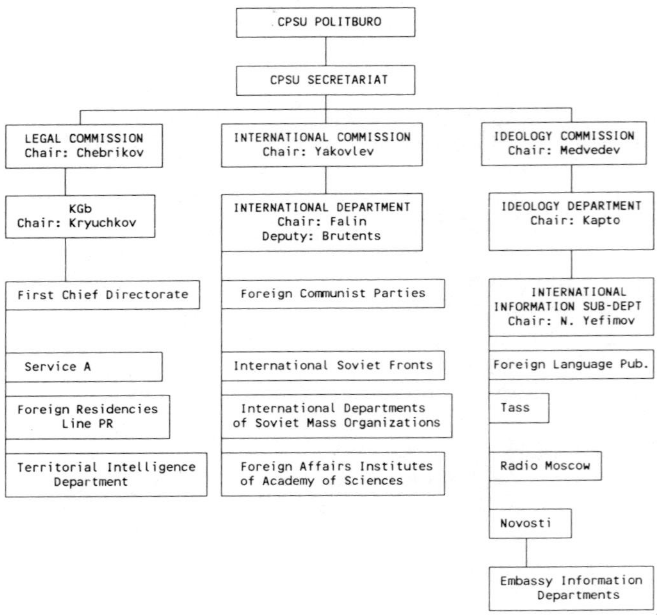

NOTES

1. *Pravda*, 1 October 1988 (in Russian) translated in *FBIS-SOV-88-191*, 3 October 1988, 40, and *Pravda*, 29 November 1988 (in Russian) *FBIS-SOV-88-229*, 29 November, 1988, 52-57. [Foreign Broadcast Information Service (FBIS), is a U.S. government translation service.]

2. *Pravda*, 1 October 1988 (in Russian) *FBIS-SOV-88-191*, 3 October 1988, 41-42.

3. Alexander G. Rahr, *A Biographic Directory of 100 Leading Soviet Officials* (Munich: Radio Liberty Research, March 1986), 221-22.

4. Konstantin Chernenko, "Topical Questions of the Party's Ideological and Mass Political Work," *Information Bulletin* (August 1983): 24. [*Information Bulletin* is a publication issued by *World Marxist Review—Problems of Peace and Socialism*.]

5. Natalya A. Dolgopolova, and Kokoshin, Andrey A., "Lessons Learned from the Destinies of Great Powers," *Kommunist* 2 (January 1988).

6. Mikhail Gorbachev, "Restructuring-People," *Novosti*, Moscow, 25 February 1987, 25.

7. Mikhail Gorbachev, "For the Sake of Preserving Human Civilization," *Novosti*, Moscow, 16 February 1987, 7.

8. Mikhail Gorbachev, "Political Report of the CPSU Central Committee to the 27th Party Congress," *Novosti*, Moscow, 25 February 1986, 12.

9. *Moscow News* appears in nine languages and is distributed throughout the world. The Russian language edition is very difficult to obtain in the Soviet Union.

10. Yegor Yakovlev, "Openness Democracy Responsibility," *Novosti*, Moscow, 1987, 10.

11. These appeared in the magazine of the Ministry of Foreign Affairs, *Mezhdunarodnaya Zhizn* (in Russian), September and October 1988, and *International Affairs* (the English language version), October and November 1988.

12. *International Affairs*, October 1988, 3.

13. Since making the speech, Vlasov was appointed head of the Novosti Press Agency and his former department has been renamed the Ideology Department. Vlasov, like many of the active measures apparatchiks, has had experience both in party and government work. From 1976 though 1980 he was counselor at the Soviet Embassy in Sweden. In 1980 he became deputy chief of the International Information Department and first deputy chief when that department was merged into the Propaganda Department. (*Izvestiya*, 31 October 1988, 4 [in Russian], translated in *FBIS-SOV-88-225*, 22 November 1988, 57.)

14. *International Affairs*, November 1988, 20.

15. Ibid., 34-36.

16. Ibid., 37 and 40.

17. Ibid., 49.

18. Ibid., 57 and 58.

19. A State Department report, *Soviet Influence Activities: A Report on Active Measures and Propaganda, 1986-87*, issued in August 1987 (23), reveals that Borovik, "was identified during a recent London libel trial as a former agent during the 1960s in Tenth Department of the KGB's Second Chief Directorate, responsible for recruiting foreign journalists. Borovik has retained close ties to the KGB through his brother-in-law, Vladimir Kryuchkov, head of the KGB's First Chief Directorate, which is responsible for all Soviet intelligence activities abroad."

20. *International Affairs*, November 1988, 60.

21. *Party Life*, issued by the Workers Party of Jamaica, (February 1982): 32. The issue contains the following instructions, "Comrades are asked to ensure that the issue of *Party Life* is kept securely and safely and, more generally, to upgrade the application of the security guidelines."

22. The Social Sciences Institute is the International Communist training apparatus in Moscow. It is also known as the Lenin School and the Higher Party School.

23. *World Marxist Review*, March 1988, 96-118.

24. *Vestnik Ministerstva Inostrannykh Del* (Bulletin of the Ministry of Foreign Affairs), 1 December 1988 (in Russian).

25. *World Marxist Review,* September 1988, 101-03.

26. *International Affairs,* November 1988, 21-22.

27. Press release, *Union of Soviet Socialist Republics Mission to the United Nations* (1 November 1988): 3.

28. The United States government, in addition to exposing specific instances of Soviet disinformation, has also published three major reports: *Active Measures: A Report on the Substance and Process of Anti-U.S. Disinformation and Propaganda Campaigns*, U.S. Department of State, August 1986; *Soviet Influence Activities; A Report on Active Measures and Propaganda, 1986-87,* U.S. Department of State, August 1987; *Soviet Active Measures in the Era of Glasnost*, United States Information Agency, June 1988.

29. *Flashes from the Trade Unions*, World Federation of Trade Unions, Prague, 8 July 1988, 4, and *The Democratic Journalist*, International Organization of Journalists, Prague (January 1989): 22-23.

30. *Political Prisoners in the U.S.A.*, National Campaign for Amnesty & Human Rights for Political Prisoners, New York (mimeographed); *Movement Support Network News*, New York (Winter 1989): 1; Press Release, *Human Rights Campaign for Political Prisoners*, 5 December 1988. All of these groups operate out of the same office in New York City and maintain the same telephone number.

31. *International Press Correspondence*, 18 June 1924 (34): 343. [*International Press Correspondence* was published in a number of languages by the Communist International. Appearing at least once a week, it printed speeches of officials of the Communist parties and reports on party activities.]

32. *International Press Correspondence*, 4 April 1925 (26): 350.

33. *The Communist International, Organ of the Executive Committee of the Communist International*, 20 September 1935, 1363.

34. A. I. Sobolev, et al., *Outline History of the Communist International* (Moscow; Progress Publishers, 1971) p. 399.

35. These are: Asian Buddhist Conference for Peace; Afro-Asian Peoples' Solidarity Organization; Berlin Conference of European Catholic; Christian Peace Conference; International Association of Democratic Lawyers; International Federation of Resistance Fighters; International Institute for Peace; International Organization of Journalists; International Radio and Television Organization; International Union of Students; Organization of Solidarity with the

People of Africa, Asia, and Latin America; Women's International Democratic Federation; World Federation of Democratic Youth; World Federation of Scientific Workers; World Federation of Teachers' Unions; World Federation of Trade Unions; and World Peace Council.

36. *The World Federation of Trade Unions, 1945-1985*, WFTU, Prague (October 1985): 25.

37. Zhukov's letter, December 2, 1982 on the letterhead of the Soviet Peace Committee (in French). A German translation appears in the West Berlin newspaper, *Die Tageszeitung*, 4 January 1983, 3.

38. James P. Cannon, *The Road to Peace* (New York: Pioneer Publishers, 1951), 8.

39. Romesh Chandra, *The Present International Situation and the Main Tasks of the Peace Movement; the Role of the World Peace Council Forms and Methods of WPC Activities in the New Conditions*, (Sofia: World Peace Council, 24-27 April 1986), 23-25 (mimeographed).

40. Zhukov's speech was mimeographed and distributed at the Sofia WPC meeting under the title *Bring the Activities of the World Peace Council in Accord with the New International Situation* (5-9).

41. The fronts on the board were Afro-Asian Peoples Solidarity Organization, Christian Peace Conference, International Association of Democratic Lawyers, Women's International Democratic Federation, and World Federation of Trade Unions. The collaborating organization was the Women's International League for Peace and Freedom.

42. The fronts now on the board are: Arab Lawyers Union, a regional affiliate of the International Association of Democratic Lawyers, Christian Peace Conference, International Association of Democratic Lawyers, Women's International Democratic Federation, World Federation of Democratic Youth, World Federation of Scientific Workers, and World Peace Council. The collaborator remains the Women's International Democratic Federation. The old and new board members were listed in the minutes of the Board for 6 September 1988 and the 7 October 1988 letter from Marek Hagmajer, president, to the Members of CONGO.

43. Mikhail Gorbachev, *Address at the U.N.*, (Moscow; *Novosti*, 1988): 30.

44. *Literaturnaya Gazeta*, 28 December 1988 (in Russian), translated in *FBIS-SOV-89-015*, 25 January 1989, 11f.

45. *Izvestiya*, 25 January 1989 (in Russian), translated in *FBIS-SOV-89-018*, 30 January 1989, 55.

46. Between December 1988 and February 1989, Segal was quoted in publication in Austria, Jamaica, Spain, Brazil, and Panama repeating the AIDS disinformation story.

47. Konstantin Nikolayev, *200 Years of the U.S. Constitution—Ideals and Reality* (Moscow: *Novosti*, 1987), 94-95.

48. Rustem Galiullin, *The CIA in Asia, Covert Operations Against India and Afghanistan* (Moscow: Progress Publishers, 1988), 111.

49. "Once Again About the CIA," (Moscow: *Novosti*, 1988), 109.

50. *Soviet News*, Soviet Embassy London, 25 January 1989, 22.

51. S.F. Alinin, B.G. Antonov, A.N. Itskov, *The Jonestown Carnage: A CIA Crime*, (Moscow: Progress Publishers, 1987).

52. Institute of Social Sciences, *Social Psychology and Propaganda* (Moscow: Progress Publishers, 1985), 260.

53. Ibid., 240

54. Michael Ledeen, and Herbert Romerstein, *Grenada Documents: An Overview and Selection*, U.S. Department of State and Department of Defense, Washington, DC, 1984, document 28.

55. Associated Press, 9 March 1989.

56. *New Times,* Moscow, No. 9, 28 February to 6 March, 1989, 11.

57. *Pravda*, 15 July 1987, translated in *Reprints from the Soviet Press*, 30 August 1987, 7, 16.

58. *CPSU: Documents and Analysis*, 4 February 1988, 10. This publication is issued weekly by *Novosti* in mimeographed form to provide Soviet Communist party documents to foreign communists.

59. *TASS*, Moscow, 20 October 1988 (in English), see *FBIS-SOV-88-204*, 21 October 1988, 52.

60. *Pravda*, 28 January 1989 (in Russian), translated in *FBIS-SOV-89-018*, 30 January 1989, 51.

61. *Pravda*, 20 February 1989 (in Russian), *FBIS-SOV-89-033*, 21 February 1989, 72-73.

5

Terrorism as a Psychological Strategy

Maurice Tugwell

> Who was he in the world who dared to show impudence
> towards this authority and who has not lost his life?
> These cursed enemies of God do not realize that their life
> and authority is approaching its limits, that
> Every one who thinks of opposing the Lord of the
> Universe, is punished by the fate of violent death.
> The chosen prophets preached and warned people about
> this from the beginning, that such is the promised
> punishment.[1]

> I inform the proud Moslem people of the world that the author of *The Satanic Verses* book, which is against Islam, the Prophet and the Koran, and all those involved in its publication who were aware of its contents are sentenced to death.[2]

INTRODUCTION

The two statements above originated within a couple of hundred miles of each other in geographical terms, but are separated in time by some 870 years. The more recent outburst by Ayatollah Khomeini is easily recognized; the earlier quotation is part of an ode written by Hasan-i Sabbah, the veritable Lord of the Mountains, founder of the Assassins, in the early 1100s.[3] Terrorism, like other forms of warfare, has ancient traditions, and past experience of its psychological strategy can assist contemporary analysis.

War is the continuation of politics by violent means. The process of peaceful persuasion gives way to physical coercion: brute force rules where sweet reason has failed. But the objective, *persuasion*, or changing the enemy's mind, has not altered, and this fact is the binding force between politics and war. In its more violent form, war may distort von Clausewitz's theory if ends are subordinated to means, and generals rather than politicians set agendas. This is quite often the case in interstate war. Revolutionary war, however, tends to adhere far more closely to the original political program, because effective revolutionaries *never* lose sight of political aims or forget that *persuasion* is the ultimate purpose of every action.

Terrorism, as a subsidiary form of revolutionary warfare, follows the same pattern in this respect. Indeed, because it is right on the line between peace and war, terrorism more than any other conflict form dodges from one side of the line to the other. One day its leaders address publics through the news media, in the manner of politicians; the next day, they blow an airliner out of the sky to underline their point. They are in the business of changing people's minds.

Purpose

This chapter examines terrorism as a psychological strategy in the context of long-term strategic planning. It does not discuss the psychology of terrorism as a sociological phenomenon, or the internal dynamics of terrorist groups.[4] To explore the way that terrorists seek to change the minds of their enemies, it is necessary to look first at warfare in general.

THE MOBILIZING TRINITY

The concept of the mobilizing trinity rests on the proposal that governments and their peoples, or revolutionaries and their followers, fight or resist so long as three essential beliefs are held. This trinity of convictions consists of:

First, a belief in something good to be promoted or defended;
Second, a belief in something evil to be destroyed or resisted;
Third, a belief in the ultimate victory of the good cause.

Shown schematically, these are the core beliefs:

INEVITABLE
VICTORY
Third Belief

GOOD
(my cause)
First Belief

EVIL
(the enemy)
Second Belief

In the traditional interstate war, each side tends to mobilize trinities along the following simple, chauvinistic but enduring lines:

Victory is
inevitable because
defeat is unthinkable.
NATIONAL SELF-CONFIDENCE

Promotion or defense
of the GOOD people.

The enemy who threatens
or resists the GOOD
must be EVIL.

In such interstate wars the weak psychological link has usually been a belief in victory. Terrible casualties contributed to the French collapse of confidence that led to the 1917 mutinies, but no French soldier went over to the German side. Likewise, the Anglo-American propaganda campaign of 1918 that contributed to the collapse of the German Army reinforced that army's growing sense of hopelessness and destroyed the inevitable victory myth. By 1940, however, the combined *revolutionary* propaganda of Nazis and communists undermined French notions of who was good and who evil. The country collapsed without suffering severe casualties.

ANTICOLONIAL WARS

During the 1945-70 anticolonial wars, relatively small, militarily insignificant nationalist groups challenged the West European imperial powers. Their mobilizing trinities were along these lines:

Nationalism is a
force whose time has come;
Our cause is NOT
to be denied.

National self-determination
is a universal right–
a shining GOOD cause.

Imperialism is EVIL,
those who defend it
deserve to die.

Against such trinities the Europeans had no psychological defense, for the Romance of Empire had given way before the Empire of Romance and almost no one in London, Paris, Lisbon, or the Hague could justify colonialism other than as a condition to be defended against unwelcome change, or in strategic or economic terms. Although diplomats and soldiers could conduct a rearguard action, there was no appropriate mobilizing trinity available to the populations of the colonial powers.

THE ASSET-TO-LIABILITY SHIFT

As a consequence, these anticolonial campaigns were conflicts where one of the parties, the colonial regime, held all the physical aces, but the insurgents had all the high psychological cards. Military inferiority sometimes led to brutal setbacks, particularly for the Vietnamese and the Algerians against the French, but in the end it was the regimes that changed their minds, not the nationalists. This often happened through the process I have called *the asset-to-liability shift*.[5]

This process works when the revolutionaries are opposed by enemies whose sole or main reason for resisting revolutionary demands is practical rather than ideological, convenience rather than survival. In Mandatory Palestine, the British resisted Zionist demands for a Jewish state because Palestine was seen as an asset. It was deemed inconvenient to vacate a valuable military base and to risk angering the Arabs. The British changed their minds not because the Jewish agency persuaded them of the superiority of their case, but because the militant terror groups, Irgun Zwai Leumi and the Stern, augmented the propaganda of the mainstream Zionists with a series of terrorist attacks. The fact that the authorities could find no way of defeating or ending these attacks made British rule seem illegitimate to international audiences, and an expensive embarrassment to British publics.

Lacking an intellectual argument, Britain weighed its options on a profit-loss basis. The Zionist campaign had robbed Palestine of its value as a base; human and financial costs were rising; the world was watching. By February 1947, Palestine was seen in London as a liability. The government handed the problem to the United Nations and the way was clear for a Zionist victory. The utility of terrorism as a means to change minds had been proven.

Over and over, the asset-to-liability shift proved to be the turning point in anticolonial wars. It failed to materialize in Malaya because the insurgents relied on communism rather than nationalism, and Britain succeeded in mobilizing the Malay population behind their own nationalist cause of Malay independence. Nor does it easily work when the territory is part of the national homeland, as with Basque terrorism or Tamil claims in Sri Lanka. In such cases, presumably, the central government is capable of depicting the terrorist campaign as a threat to basic national values and unity: a defensive trinity is created.

VIETNAM

During the Vietnam War the U.S. government found itself in the same trap that Britain, France, and others had fallen into during their anticolonial wars. The mobilizing trinity was weak. To be sure, the United States had a record of military success that provided initial confidence that it would prevail; in the 1960s, communists were generally seen as bad guys; and the cause of protecting democracy in the south was fine. But there was no direct threat to the USA, and the war was a sideshow rather than a central commitment.

This allowed the relatively weak North Vietnamese to gain the psychological upper hand and defeat their militarily stronger enemy. By portraying the National Liberation Front as nationalist and independent, they undermined the evil enemy part of the trinity; by depicting the U.S. military effort as genocide and institutionalized brutality, they weakened the good image; and by denying the United States a quick victory, through Tet, they eroded hope that victory could ever be won.[6] In the United States there were really two end-the-war movements: an intellectual and patriotic lobby that accepted that the war should be ended because victory was not in sight, and a disloyal, radical movement that painted the North Vietnamese as the good guys and the Americans as the evil enemy. The second group extended the conflict into the United States in full revolutionary manner.

Whereas the Europeans sought to retain their colonies for material reasons that ceased to appeal once the price was too high, the Americans had fought in Vietnam for the *principle* of freedom. The impact of the asset-to-liability shift was therefore particularly harsh, because the asset that was abandoned was a commitment to defend freedom. (Although the circumstances are quite different, it is to be wondered what psychological effect the recent Soviet withdrawal from Afghanistan will have on generations brought up to believe in historical determinism.)

IDEOLOGICAL TRINITIES

The trinities that undermined French cohesion in 1940 and the United States' will to prevail in Vietnam in the late 1960s were ideological, such as these:

The Invincible
National Socialist Dynamic
of the 1000 Year Reich

The NAZI Party THE MASTER RACE (Perverted German Nationalism)	Inferior Races-- Jews, Slavs, etc. EVIL enough to die

and this:

HISTORICAL
DETERMINISM
Inevitable Victory

THE EXPLOITED CLASS The CPSU International Communism (THE FUTURE)	THE EXPLOITING CLASS Capitalism Imperialism (THE PAST)

and, in the same category though religious rather than ideological:

THE WILL OF GOD
(Messianic Belief)
makes Victory Inevitable

FAITH in the Righteous Cause.	Insofar as Heretics offend, they are EVIL in the eyes of God.

These three are typical of the urgent, generally intolerant, mobilizing trinities that have to some extent superseded nationalism as prime legitimizing forces for conflict. The last illustration applies equally to the terrorism of the eleventh and twelfth century Assassins and to the terrorism of contemporary militant Shia Islam.

Few trinities are as pure as the examples shown. Ideology is overlaid with nationalism, religion with anticolonialism, and so on. We can imagine the type of trinity that mobilizes the extreme, left-wing elements of the Palestine terrorist movement, such as the Popular Front for the Liberation of Palestine (General Command), the Democratic Front, and the Abu Nidal organization:

Historical Determinism,
the Will of God, and Palestinian
National Destiny

The Palestinian Cause
is GOOD because the
land is ours.

ISRAEL, IMPERIALISM,
THE UNITED STATES and
WEST. All who oppose.

Whatever their content, such trilogies provide today's terrorists with *certainty*. They are doing what is right; their victims deserve to die; it is not a waste of lives and effort because their cause is bound to triumph in the end. For their enemies, the psychological situation is rather different.

THE WESTERN TARGET STATE

Democracy can never be a rigid ideology and the concept of permanent mobilization, so typical of the theocracy and single-party state, is impossible. A healthy liberal democracy may enjoy a consensus on certain key matters—the rule of law, the importance of human rights, the value of the democratic system—but it takes an unambiguous threat to conjure up the climate of opinion that will support full war mobilization or any sustained military or quasi-military commitment. Without such a commitment, the very strengths and ideals of democracy can render it especially vulnerable to terrorism, for it is open, it is limited in its response by law and concern for individuals, it has a free media, and it will inevitably contain domestic groups and individuals whose sympathies lie with the terrorist cause if not with their methods.

In the world today, the West is the likely target of terrorism of the far left, whether allied to the USSR, Cuba, militant Islam, violent

Palestinians, or, more selectively, minority nationalist causes that are seen as being denied by a Western country, such as the Basque, Irish, and Puerto Rican. As the leader of the Western Alliance, the United States is always a prime target for anti-NATO, anticapitalist, and anti-imperialist terrorism. As chief supporter of Israel, the United States occupies a similar position in the Middle East.

This does not mean that most terrorism occurs in Western countries. In 1988 Pakistan was target Number One, and countries such as Lebanon, Syria, South Korea, and of course Sri Lanka, Colombia, and Peru have endured massive attacks. However, in the strategic sense it may be true that the West is the prime target.

Given such priorities, it is remarkable that the United States itself, as opposed to assets abroad, has been so free from attack. Is this due to the efficiency of the FBI and the difficulties facing terrorists in establishing bases? Or does it reflect concern not to awaken the giant, not to create a trinity?

TERRORIST PERSUASION IN ACTION

Hasan-i Sabbah and his Assassins never tried to conquer Persia or Syria, being content to govern only their mountain bases and to spread their creed, called "The New Preaching," throughout the Islamic world. What they demanded from the Seljuk authorities was respect for their creed and freedom of action for their missionaries, so provincial governors who stood in the way and Sunni clerics who spoke out against the New Preaching were dealt with.

The usual penalty was death; the method, the Assassin's dagger. The intimidatory power of the Assassins rested on their reputation of always killing their chosen victims, however long it took. Precisely because of this reputation, they could afford to forgive some targets if the propaganda payoff justified it. A famous Sunni theologian, Fakhr al-Din was an effective critic of the New Preaching. Killing him would shut him up but possibly make him a martyr in the Sunni cause. So an Assassin overpowered him, threatened him with the dagger, and offered him a generous annuity in gold. At once terrified and tempted, the teacher agreed to change his mind. Later, when, a student commented on his altered views, Fakhr al-Din replied, "It is not advisable to curse the Ismailis, for they have both weighty and trenchant arguments."[7] The great man had been compromised and neutralized. Today, amid shouts of defiance, book publishers are quietly cancelling pocketbook editions and translations of *The Satanic Verses* and stores have pulled the book from their shelves. Salman Rushdie, the offending author, has half apologized, half changed his mind, but the death sentence stands. When, in the West, or anywhere else for that matter, will another book be written or published if there is even the slightest risk of it creating the same response? Few will admit that their minds have been changed, or that the principle of free expression so long cherished as an asset has quietly been discarded as a liability.

TERRORIST PSYCHOLOGICAL STRATEGY

These two examples of terrorism in action provide some clues to its psychological strategy. During the past two years the nature of international terrorism has changed considerably; the shift is away from publicity-seeking dramas, from lengthy hostage situations designed to force governments to make concessions such as the release of political prisoners, and towards low-intensity wars of attrition apparently intended to have strategic rather than tactical consequences. This revised strategy may be reduced to these guidelines, seen from the terrorist point of view:

- Set limited objectives. These can be incremental, but one step at a time is the rule.
- Achieve objectives by persuasion.
- Use terror to persuade.
- To terrorize, demonstrate ruthless effectiveness, create uncertainty, and let no one feel safe.
- Avoid creating a Pearl Harbor reaction to your actions because this could produce a mobilizing trinity in the target nation(s).
- Instead, maintain a remorseless campaign at a pace that frustrates, angers, confuses, divides, humiliates, and *conditions* target audiences, without quite awakening them.
- Deny the enemy victories. You must be seen as invincible; the target must feel helpless.
- Condition the enemy to accept warfare on your terms, such as safe haven in pro-terrorist states, the use of hostages to discourage effective reactions, the use of fronts and intermediaries to conduct negotiations, or make propaganda in parallel with terrorism. Conditioning leads to compromise and compromise leads to surrender.
- If the immediate objective is modest and your campaign generates high costs, the asset-to-liability shift should apply. Take your prize, then make the next demand.
- Discourage any incipient trinity in the target society by attacking all three components: Demonstrate that terrorism can never be defeated; undermine target self-esteem by guilt transfer[8]; promote the terrorist cause by propaganda, using third parties.
- Switch from terror to nonviolent politics and back again whenever appropriate, taking credit for the first and blaming the target for the second.

- Reward previously hostile states that make deals by excluding their citizens and assets from attack. But, establish bases in such states and let it be understood that any interference will be punished severely.
- Remember throughout that you are changing the mind of your enemy. Proposed actions must be examined against this criterion. Some actions can rebound.

TERRORISM AND EAST-WEST RELATIONS

Mikhail Gorbachev's New Preaching (Thinking) evidently promotes nonviolent persuasion to pride of place in Soviet international policy. Direct Soviet aggression is highly unlikely; some forms of indirect revolutionary warfare seem to still be in season, as are subversion, propaganda and ideological conflict, and intelligence warfare. Soviet support for anti-Western terrorism appears to be at a crossroads.

It seems quite possible that the USSR will adopt a dual policy over international terrorism, designed to allow Moscow to occupy the moral high ground by denouncing terrorism, hosting international conferences on the subject, and leading debates in the U.N. The aim would be to distance the USSR from terror sponsorship while associating such behavior with the United States (the Contras, 1960s CIA assassination plots, etc.), and to redefine terrorism in Third World terms as *resistance* to communist-led revolution.

At the same time it would be possible, covertly, to encourage anti-Western terrorism by groups and their sponsor countries outside the Eastern Bloc—particularly the Islamic Jihad and the Abu Nidal organization supported by Libya, Syria, and Iran. West European groups, such as the Red Army Faction, Action Direct, and the Red Brigades, appear to react to propaganda direction and their future campaigns could be backed as necessary from Middle Eastern sources. It ought to be recognized that Soviet intelligence has achieved a dominating position in the West and Third World, permitting Moscow to do pretty much as it likes without being caught with a smoking gun.

An indication of Moscow's real attitude toward terrorism was supplied by an August 22, 1988, Moscow *Domestic Service* newscast on General Zia ul-Huq's death:

> In a human way, Zia ul-Huq's tragic fate is understandable. But in a human way it was also to be expected. For the greater part of his eleven years in power, the General was a dictator, and sometimes a very cruel one. And here President al-Sadat springs clearly to mind. The circumstances of their deaths may not be the same, but their biographies as faithful servants not in the interests of their own people, but rather those of overseas protectors,

> are almost identical. Who is to say? Perhaps it is precisely here that lie hidden the reasons that put such a sudden end to the President's last flight.[9]

The message is abundantly clear: Those who oppose Moscow's New Preaching deserve to die, and whether they die at the hands of the KGB or domestic terrorists really does not matter. "Who was he in the world who dared to show impudence towards this authority and who has not lost his life?"

In addition, Soviet psychological gains from anti-Western terrorism—which do not necessarily coincide with the objectives of the terrorists groups—could be as follows:

- Diversion of resources from defense (NATO, etc.) to counterterrorism;
- Divisions within the Western Alliance (over response, responsibility, etc.), and within target nations (loss of government legitimacy, blame for failures, resentment at protective measures);
- Guilt and hostility in Western societies toward the outside world, which is seen as ungrateful and aggressive, leading to public demands that governments withdraw from their international leadership roles and mind internal problems;
- A state of fear and uncertainty hampering decision making and policy execution.
- Ultimately, the discrediting of the free, open society as being incompatible with security; freedom, formerly an asset, is seen as a liability.

No one supposes that these possible objectives are likely to be achieved in the near future, or easily. But given the fact that a covert Soviet reinvestment in antiWestern terror would be virtually cost-free, the chance of achieving even some of these psychological benefits may be too tempting to resist.

Indeed, as though to warn the West not to expect too much, the Soviet-controlled World Peace Council, which has traditionally confined its applause to wars of national liberation and not to terrorists, included in the March 1989 issue of its *Peace Courier* an article by Niall Farrell, a member of the National Executive Committee, Communist Party of Ireland, supporting the IRA offensive. Farrell's piece was designed to deprive the British government of public confidence in its counterterrorist measures, thus giving the IRA a freer hand. Propaganda to delegitimize any effective counterterrorist individual, unit, or method has been a constant feature of IRA terrorism since 1970.

CONCLUSION

Although we naturally tend to fear acts of terrorism because of the damage they can cause to our bodies and property, the terrorist threat to society as a whole, and to nations, is essentially psychological. Like other types of combatants, terrorists try to change the minds of their enemies. They operate from strong beliefs in their own righteousness, in the evil of their enemies, and in eventual victory. Since these beliefs are essential to their continued operations, authorities ought to direct as much attention to neutralizing hostile trinities and to changing terrorists' minds as to defeating the violent plans that flow from them, and they should consider the psychological impact of terrorism on their own populations as seriously as they would consider the physical impact.

NOTES

1. "Ode by Hasan in Praise of Three Fidawis," translated W. Ivanov, *Journal of the Bombay Branch of the Royal Asiatic Society*, 14 (1938): 63-72, lines 23-26.

2. Ayatollah Khomeini, quoted *Daily Telegraph* (London), 15 February 1989, 1.

3. On the Assassins, see Bernard Lewis, *The Assassins: A Radical Sect in Islam* (London: Weidenfeld & Nicolson, 1967).

4. On this, see Maurice Tugwell, "Terrorism and Propaganda: Problem and Response," P. Wilkinson and A.M. Stewart (eds.), *Contemporary Research on Terrorism* (Aberdeen: Aberdeen University Press, 1987).

5. See Maurice Tugwell, "Revolutionary Propaganda and Possible Counter-Measures," unpublished Ph.D. thesis, University of London, 1979.

6. See Guenter Lewy, "Deception and Revolutionary War in Vietnam," in David Charters and Maurice Tugwell (eds.), *Deception Operations: Studies in the East-West Context* (London, New York: Pergamon-Brassey's, forthcoming).

7. Lewis, 75-76.

8. See Maurice Tugwell, "Guilt Transfer," in David C. Rapoport and Yonah Alexander (eds.), *The Morality of Terrorism* (New York: Pergamon, 1982), 275-89.

9. Quoted in *Nightwatch* (Arlington, VA) 3, no. 10 (October 1988): 9.

6

Soviet Diplomacy and Public Opinion: The Case of West Germany

Josef Joffe

West Germany has been a classical locus of Soviet public diplomacy. The intent has always been to influence public and political opinion in favor of Soviet foreign policy objectives, and the attempt is as old as the Federal Republic itself.

The most prominent instance, though by now almost four decades in the past, was Stalin's notorious March 1952 offer of a reunified, but neutral Germany. The impetus for this sudden but not quite unexpected foray was the impending rearmament of the Federal Republic within the framework of the European Defense Community (EDC). Providing for the addition of German military power to the Western coalition, these talks with the three Western powers were on the verge of consummation by the spring of 1952. Hence Stalin's diplomatic offensive, which was obviously designed to torpedo the process by dangling neutralization-cum-reunification before the West Germans. The gambit failed, and in 1955, the Federal Republic joined NATO as a sovereign member.

Since then, the Federal Republic has been a routine target of Soviet pressures and blandishments, and the occasion has always been an impending defense choice in the framework of NATO. In the late 1950s, Moscow unleashed the most massive campaign to date—when the Adenauer government decided to equip the *Bundeswehr* with delivery vehicles (artillery, aircraft) for U.S. tactical nuclear weapons. Including threats of nuclear incineration in case of war, the campaign was launched in earnest in 1958, and it did not abate until after the Berlin Wall was built in 1961. Indeed, many thoughtful observers have surmised that the Berlin Ultimatum, flung down by Nikita Khrushchev in November of 1958 was above all a response to Adenauer's nuclear choice as a most forceful way of pressuring the West Germans out of their nuclear commitment.[1]

It was no accident that the next campaign of threats and domestic interference came in the late 1970s and early 1980s, when the Federal Republic served first as the main potential deployment area of the so-called neutron bomb and then as the very pillar of the *Nachrüstung*, the stationing of Pershing II and cruise missiles in Western Europe. A burgeoning peace movement, mobilizing hundreds of thousands of demonstrators in the streets and squares of West German cities, offered a natural target for Soviet public and not-so-public diplomacy.

The peace movement was a motley crowd, uniting old-style Leninists and starry-eyed idealists, pastors and pacifists, Greens, and Social Democrats. Yet although Moscow's surrogate in West Germany, the *Deutsche Kommunistische Partei* (DKP), ardently tried to gain control over the movement, it never did. One of the DKP's main contributions to the peace movement was logistics, as this author found out when he researched the organization of the huge 1981 antinuclear demonstration in Bonn. The protesters were carted to Bonn from all over West Germany in chartered trains. Inquiries as to the organizers at the charter office of the *Bundesbahn* (Federal Railway) resulted in telling replies such as Clara Zetkin Women's Group, Rosa Luxemburg Bookstore, "Solidarity With the Third World"—all front organizations for the DKP. Yet Moscow-style communists were but one part of the crowd. They mingled and collaborated with militant Protestants, left-wing Social Democrats, Greens, and nonaffiliated protesters. The Soviet Embassy in Bonn as well as the news services TASS and Novosti provided plenty of propaganda fire. Still, the peace movement never obeyed a single voice, let alone a single controller. Not even the Social Democrats (SPD) who hoped to coast to power on the antinuclear vehicle ever managed to gain control over the movement.

The peace movement quickly picked up on Soviet propaganda cues such as the Pershing II being a first-strike weapon. The movement also amplified Soviet threats such as West Germany becoming a foreordained target of Soviet nuclear strikes in case of INF deployments. But it is important to remember that the peace movement of the 1980s, unlike the peace movement of the 1950s, where Moscow's communist surrogates played a crucial role, was primarily an indigenous movement that served Soviet purposes willy-nilly rather than by design. This is all the more important to remember as we turn our attention to the contemporary West German scene.

To put it bluntly, Soviet psychological warfare may no longer be as important in West Germany as it was in the past. For reasons which are not (or at least not directly) of Moscow's making, West German public opinion has been transformed in ways that must be highly encouraging to Soviet policymakers.

What has not changed are the basics: the population's attachment to NATO, to the U.S. military presence, and to the *Bundeswehr* (Federal Armed Forces). Approval rates on all three items have remained where they used to be in the past: in the 70 to 80 percent range.

- "Should the Federal Republic remain a member of NATO or not?" Sixty-eight percent (1987) say yes, 10 percent say no—virtually no change since 1971 (71 percent say yes).[2] Another survey shows 75 percent in favor of staying in NATO (1987).[3]

- "Are you for or against the withdrawal/partial withdrawal of U.S. forces in the Federal Republic?" Seventy-five percent are against any withdrawal, only 23 percent are in favor. This is a higher approval rate than in the 1970s when only 60 percent were against any kind of withdrawal.[4]

- "Does the Bundeswehr make the peace more/less secure?" In 1987, 85 percent opted for "more secure," up from 69 percent in 1974.[5]

So much for continuity. On other items, the transformation of public opinion in West Germany seems nothing if not dramatic—no matter which dimension is chosen. Willy-nilly, they all favor the fundamental objectives of Soviet foreign policy.

1. *Attitudes Toward the Soviet Union*: Seventy percent of the West German public displayed a very/somewhat favorable opinion toward the Soviet Union in March 1988. (In France, it was 40 percent; in Britain, it was 59 percent.) The change has been both rapid and pronounced, as only 47 percent of West Germans took such a benign view of the Soviet Union in September of 1987. At that time, the corresponding French figure was 30 percent, and the British one was 48 percent, which suggests a more rapid shift in Germany than in Britain and France.[6]

By mid-1988, a staggering 71 percent of the West German population had a favorable opinion of Soviet General-Secretary Mikhail Gorbachev—which is a rating normally accorded only to the West German president, a largely ceremonial figure above the fray of political battle. By contrast, only 41 percent took such a favorable view of President Reagan.[7] Distrust toward the East declined from 58 percent (1982) to 25 percent (1988). Forty-nine percent believe that Soviet armaments are defensive, hence there was no need to feel threatened; only one-quarter of the population believed that Soviet military might serves as an instrument of pressure, designed to achieve hegemony over all of Europe.[8]

2. *Attitudes Toward Nuclear Deterrence*: The very premise of nuclear deterrence—that it has kept the peace in Europe—is apparently no longer shared by a majority of West Germans. The query was phrased as follows: "Some people say that more than 40 years of peace in Europe is owed to nuclear weapons. Do you agree, or not?" More

than one-half of the sample did not agree, only 30 percent did.[9] Almost six out of ten West Germans believed in 1987 that deterrence can be achieved through nonnuclear means; only two out of ten still believe in the necessity of *nuclear* deterrence. Compared to 1983, this is a shift of 11 and 7 percents, respectively.[10]

3. *Attitudes Toward Nuclear Weapons*: "Would you like to see all nuclear weapons withdrawn from Europe?" A startling figure, 79 percent of West German respondents, replied yes, and only 10 percent said no.[11] The rejection of new nuclear weapons was less dramatic, but still overwhelming; 63 percent are against the "modernization of short-range nuclear missiles," while only 14 percent are in favor.[12]

4. *Attitudes Toward Unilateral Disarmament*: Those West Germans who favor unilateral disarmament as a road to détente and peace have risen from 33 percent in 1981 to 48 percent in 1986, while the number of those who believe that this creates insecurity and vulnerability to blackmail has fallen analogously (from 47 to 35 percent).[13] One-half of the respondents said that "the West ought to begin with disarmament," a shift of 15 percentage points since 1982; 28 percent opposed this course, down from 41 percent in 1982.[14]

5. *Attitudes Toward the Political Value of Nuclear Weapons*: Just slightly more than two out of ten West Germans believed (1987) that the Federal Republic will be vulnerable to Soviet pressure "once there are no more nuclear missiles in our country." Almost six out of ten rejected this likelihood.[15]

6. *Attitudes on Substituting Conventional for Nuclear Military Power*: "Assume that all nuclear medium-range missiles are withdrawn from Europe. Are you for spending more money on a purely conventional defense, i.e., one without nuclear weapons?" In 1988, almost seven out of ten replied no to this query; only 16 percent were willing to devote more money to a purely conventional defense.[16]

What does this survey of contemporary public opinion signify? On the one hand, the foundations still seem to be intact, as evidenced by the high approval rates for abstract political goods such as NATO, the U.S. military presence, and the West German armed forces. On the other hand, when the questions move from the abstract to the specific, when they seek to plumb the price the West German population is willing to pay for such good things as NATO and U.S. protection, then approval rates plummet dramatically. There is, in other words, the profound erosion of the acceptance of nuclear weapons and of nuclear deterrence as such, but without the willingness to compensate for nuclear withdrawal with a greater conventional effort. These changes go hand in hand with a drastic decline of threat perceptions that served to legitimize the acceptance of nuclear weapons and of a heavy defense burden in general. The message of public opinion is: We still like NATO and the U.S. presence, but we are dubious about the *means* necessary to buttress both, that is, large-scale maneuvers on home ground, low-level training flights, modernization of nuclear weapons, increased conventional defense spending, indeed, nuclear deterrence altogether. Add to that another important facet of public opinion, a 44 percent approval rate for a "policy of neutrality between the great powers in East and West,"[17] and the basic message is underlined once more: We want good things like protection, but we don't want to pay for them as much as we once did.

Given some key continuities of Soviet *Westpolitik*, such a panorama of public opinion offers an almost ideal vista. At a maximum, Soviet objectives have been to deny to the West the enormous military and political potential the Federal Republic represents in the overall balance of power—hence the repetitive attempts from 1948 (the Berlin Blockade) to 1955 (the FRG's integration into NATO) to forestall West Germany's inclusion into the anti-Soviet coalition organized by the United States.

Thereafter, with the FRG's integration into the West a *fait accompli*, Soviet objectives were scaled down to deny or devalue the *nuclear* asset the Federal Republic represents in the scheme of Western defense. Hence, the all-out Soviet drive in the late 1950s to prevent or reverse West Germany's decision to go nuclear (with tactical weapons under U.S. control). Hence, the Soviet offensive in 1977-78, targeted in the main on the Netherlands and the Federal Republic, was designed to forestall the deployment of the neutron bomb in Western Europe. And hence, from 1979 to 1983, the campaign of threats and pressures to derail the impending deployment of Pershing II and cruise missiles; here again, the main target was the Federal Republic, the cornerstone of the entire enterprise. Had Bonn reneged, Holland and Belgium certainly would have followed suit, perhaps Italy, too, leaving only Britain to accept its contingent of cruise missiles.

For almost 40 years, Soviet objectives toward West Germany were frustrated. The FRG refused to be baited by Stalin's reunification-cum-neutralization offer in 1952. Bonn accepted rearmament under Western auspices as well as NATO membership in 1955. In the late 1950s, Adenauer managed to realize his unpopular decision to equip the West German army with nuclear delivery vehicles, as did his successor Helmut Kohl with regard to the Pershing IIs and the cruise missiles.

Today, however, Mikhail Gorbachev seems to have triumphed where all his predecessors had failed. In February 1989, Helmut Kohl refused to consent to the modernization of NATO's short-range forces (SNF) with 88 *Lance* launchers, postponing a decision until 1991, by which time conventional disarmament in the context of the Vienna CFE talks might have rendered such modernization unnecessary—or at least out of phase with the spirit of the time. As the leftist *Liberation* (Paris) put it tersely on February 13, "Helmet Kohl reads the opinion polls." With 63 percent of the populace against modernization and a staggering 79 percent in favor of total denuclearization (see above), the chancellor no longer seemed to have the stomach for a replay of the *Nachrüstung* battle of the early 1980s. By itself, as the experience of the early 1980s showed, public opinion might not have been decisive in Kohl's calculation. Yet other, more weighty parameters had also turned against the man who had so handily bested the foes of nuclear weapons in the early 1980s.

For one, Kohl's foreign minister and junior coalition partner, Hans-Dietrich Genscher (who had valiantly fought for the INF deployment) was now dead set against modernization. Genscher and his Free Democrats, moreover, hold the balance of power in the Kohl coalition; without the FDP, the Christian Democrats would have no chance to form another government after the 1990 national election. Second, West Germany's Conservatives—Mr. Kohl's own cohorts—have changed. Once the very guardians of the German-American relationship and of NATO's nuclear orthodoxy, key Christian Democrats are now on record for the drastic reduction, if not total elimination, of nuclear battlefield weapons, and none of them is a fervid supporter of SNF modernization. Third, the Russians have changed. Whereas Brezhnev and successors had piled nuclear weapon on nuclear weapon in the late 1970s and early 1980s, Mikhail Gorbachev offered wholesale disarmament, of both the nuclear and conventional variety. The Cold War, in short, seemed to be over for good, at least in the minds of the German populace (see public opinion data above), the Social Democrats, the Greens, most of the Free Democrats, and influential segments of published opinion, both print and electronic.

In other words, what Soviet political pressures and arms investment could never gain, Mr. Gorbachev's strategy of depressurization—a combination of disarmament offers, domestic reform, and diplomatic flexibility—seems to have achieved: a Federal Republic that is no longer just the pillar, but is also a question mark of Western defense. In a way, the psychological profile of the Federal Republic at the threshold of the 1990s is virtually tailor-made for the grand strategy of the Soviet Union.

On the one hand, West Germany's continuing attachment to NATO and the U.S. military presence must be seen as a blessing in disguise for the Soviet Union. The Federal Republic's continuing integration in NATO provides a measure of (Western) supervision over the one nation in Europe whose power, both actual and potential, the Soviets have always had reasons to fear. Western military forces in the Federal Republic not only assure that measure of supervision; they also legitimize the presence of Soviet forces in Central and Eastern Europe.

On the other hand, the Federal Republic, traditionally the "continental sword" and forward bastion of U.S. power in Europe, is moving—albeit slowly and unconsciously—into a more self-assertive position. With regard to nuclear modernization, Bonn is claiming—and so far, has gained—a veto power over Alliance choices. With regard to the enormous conventional and nuclear burdens the FRG has willingly carried for thirty-odd years, Bonn is similarly claiming a decisive voice in NATO councils, and successfully so, hence the cancellation of the

1989 *Reforger* maneuvers and the large-scale reduction of low-level training flights demanded by Bonn and which will surely be respected by the Alliance. All together, the FRG is—again both unconsciously and fitfully—moving into a role in the 1990s where the country in the middle might once more take a page out of Germany's grand strategy of the past: though anchored in the West (via NATO and the European Community), the Federal Republic might try to use its strategic position in the center to determine the pace and structure of East-West relations in Europe.

From a Soviet point of view, such a role, that is, the "Germanization" of East-West relations in Europe, simply might obviate the cruder means of pressure, public diplomacy, or domestic interference. The INF agreement of 1987, removing longer-range weapons and leaving in place short-range systems destined to explode on German soil, has sharpened West Germany's sense of endemic exposure as both glacis and stationary nuclear weapons carrier of the Alliance. The trend, as a result, is in the direction of *more* denuclearization and the progressive demilitarization of East-West relations in Europe. This is precisely, and again willy-nilly, the ancient objective of Soviet grand strategy. And the underlying reason is as old as alliance politics itself. If a nation feels that its *supply* of security is dwindling, it may try to reestablish equilibrium by diminishing its *demand* for security. The appropriate strategy is a policy of partial propitiation. Such a policy would seek to lessen the incentives the Soviet Union might have to pressure and threaten the Federal Republic. The means are diplomatic accommodation and economic/financial side payments, which happen to be the very essence of *Ostpolitik*.

Twenty years ago, when Willy Brandt embarked on the New Ostpolitik, this policy still met with fierce domestic opposition on the part of the Christian Democrats and large segments of public opinion. Today, the (largely still implicit) logic of West German grand strategy and the trends of domestic opinion have combined to smooth over ancient cleavages; there is what one might call the reconstitution of the national security consensus on the left. On the great issues of nuclear weapons, the demilitarization of East-West relations and quickening cooperation with the Soviet Union and its allies, Christian Democrats and Social Democrats, are no longer separated by deep cleavages but merely by degrees.

From a Soviet point of view, this is the best of all possible worlds. Domestically, there is a favorable public opinion spectrum (again, indigenously formed rather than created from the outside) which the Prague journal *Problems of Peace and Socialism* has described as follows: "To the main forces shaping today's social evolution belong the democratic mass movements (ecologists, Greens, peace organizations, womens' groups, opponents of nuclear power, grass-roots initiatives, etc.). Those movements are objectively opposed to the policies of the most reactionary circles of Imperialism, and they all flow into the general stream of the struggle for peace and social progress. Therefore, the Communist and Workers' Parties attach so much importance to cooperation and common action with these movements."[18]

Diplomatically, the opportunities are no less promising. To quote from *Problems of Peace and Socialism* again: "In the face of the masses longing for peace, the large Social Democratic parties of the FRG and Denmark have changed their foreign policy ideas in a positive sense."[19] Moreover, and here the journal relates Soviet actions to Western changes in attitudes: "The USSR's tireless effort in favor of peace and disarmament has achieved that the thesis of 'equi-responsibility of the superpowers' for international tensions has clearly lost its hold. In Austria and in the FRG, in Sweden and Denmark, more and more people have realized that Washington is the main culprit when it comes to the danger of war."[20]

That may be an exaggerated rendering of events, but it is certainly true that in West Germany (see the public opinion data quoted above) attachment to NATO and the U.S. military presence goes hand in hand with a rather benign vision of the Soviet Union. Moreover, in the Federal Republic Gorbachev's Russia faces a country that has become both more self-assertive and less certain about its identity and vocation—no longer confident in the protection afforded by its allies yet unwilling to make up for the shortfall and thus aiming for stable accommodation with the Soviet Union. As the FRG fitfully loosens the bonds of alliance, its weight in Western councils—already large by virtue of its power and position—has been increasing because neither the United States nor the European community can afford to alienate the country that is the very linchpin of the Western communities.

The Federal Republic of the 1990s is thus at the heart of two interlocking special relationships. It is the most important European member of the Western alliance and thus the *primus inter pares* whose voice will be heeded, if only for fear of multiplying ancient *incertitudes allemandes*. At the same time, and precisely because of its weight in the West, the FRG is the prime interlocutor of the Soviet Union—as a partner who is a vulnerable *demandeur* in many respects (with regard to lowering the military threat from the East and enhancing the ties to the GDR) and thus ready to act as intermediary for Soviet and East European interests in the West. For a Soviet Union that will always have to guard against hostile coalitions in the West, this is truly the best of all possible worlds: better than a Federal Republic totally integrated in, or totally alienated from the West. And this explains why the cruder means of pressure, manipulation, and interference are no longer as tempting as they were in the past.

NOTES

1. Significantly, part of the Berlin Ultimatum, which sought to turn West Berlin into a free city, was the demand for the withdrawal of both German states from their alliance systems and their complete disarmament, apart from those forces necessary to maintain domestic law and order.

2. Elizabeth Noelle-Neumann, *Die NATO-Experten und das Publikum: Das Problem der Akzeptanz von Verteidigungspolitik* (Allensbach: Institut für Demoskopie, 1988), 12. This is a most useful and up-to-date collection of public opinion data on West German defense and foreign policy.

3. *Meinungs—und Haltungslage zur Sicherheitspolitik und Bundeswehr* (Informations—und Pressestab, Bundesministerium der Verteidigung, 26 November 1987, mimeographed), Table A1.

4. Ibid., Table A7.

5. Ibid., Table A2.

6. USIS, *Research Memorandum*, 4 April 1988, 6.

7. Elizabeth Noelle-Neumann, *Die NATO-Experten und das Publikum,* Annex, Table A2. The gist of the data collated in this paper can be found in Noelle-Neumann, "Wenn das Gefühl der Bedrohung schwindet," *Frankfurter Allgemeine Zeitung*, 22 July 1988.

8. Ibid., Tables A3 and A4.

9. Ibid., 6, Table 1.

10. Ibid., Table A12.

11. Ibid., Table A6 (data gathered in August and September of 1987).

12. Ibid., Table A7. There is a curious inconsistency here between those 79 percent who favor total denuclearization and those 63 percent who reject modernization. Theoretically, everybody opposed to nuclear weapons as such ought to be opposed to nuclear weapons modernization, too. Yet the opposition to the latter amounts to only 63 percent. The explanation must be sought in the cues provided by the question. The withdrawal question contained both "nuclear weapons" and "Europe" as cues. The modernization question had a more

complicated wording: "After the elimination of INF, NATO will have only short-range missiles for deterrence, which, however, will be obsolete soon. Should these be modernized or not?" This wording contained a positive cue, "NATO," a semi-positive one, "deterrence," and a suggestive one, the approaching obsolescence of remaining nuclear systems. That kind of formulation, it seems, is good for a differential of at least 16 points.

13. Ibid., Table A10.

14. Ibid., Table A11.

15. Ibid., Table A14. The exact figures are 22 and 57 percent.

16. Ibid., 8, Table 2.

17. Ibid., 13, Table 4. In 1980, only 31 percent favored such policy.

18. The monthly has been published in Prague since 1958; it appears in 42 languages; its board includes 70 Communist and Socialist parties. Traditionally, the editor has always been a Russian. The quote is taken from the German language edition published in East Berlin. "Gleichberechtigte verbündete, ehrliche Partner: Die Kommunisten und die neuen sozialen Bewegungen," *Probleme des Friedens und des Sozialismus*, no. 12 (1986): 1642. This article is based on the minutes of a roundtable discussion with high-level members of the CP's of Austria, Belgium, Denmark, Portugal, Switzerland, and Sweden. The West German representative hailed from the "Institute for Marxist Studies and Research" in Frankfurt. Discussion leader was the editor of the Journal, Sergei; Zukassov of the CPSU.

19. Ibid., 1650.

20. Ibid., 1651.

7

Communist Regimes' Psychological Warfare Against Their Societies: The Case of Poland

Teresa Rakowska-Harmstone

The concept of psychological warfare is commonly used to describe foreign policy activities directed against an adversary either during the war or in peacetime in order to generate desired political outcomes. This chapter is concerned with psychological warfare as an instrument of strategic, long-range social conditioning in domestic politics, and is thus a novel venture. In the case of communist regimes in Eastern Europe, an effort on the part of a foreign-imposed regime to alter perception and manipulate cognitions and the direct behavior of its newly acquired citizens implies a magnitude, conflict, and intensity greater than the standard socialization effort by political establishments elsewhere. Having successfully completed the necessary first phase of physical annihilation of all organized resistance and its leaders, but still operating in an almost totally hostile environment, new regimes continue the warfare by psychological means, in the struggle for the minds of the people, in order, first, to maintain power and second, to generate legitimacy.

It is the condition of continued warfare that differentiates an imported regime's effort to influence the thinking and actions of its people from customary propaganda activities of other governments, even as it shares with the latter such characteristics as the deliberate and manipulative nature of the effort, the full use of modern communications techniques, and the latest discoveries of social psychology. In common with other communist regimes, it also employs a massive state/party indoctrination apparatus, the work of which is always given top priority and is applied on a mass and massive scale. A perception of warfare against class enemies, moreover, is an integral part of communist *Weltanschauung,* and permeates the domestic politics of all Marxist-Leninist regimes, especially their attempt to create a "new socialist man."

Conditioning in support of system maintenance includes a broad spectrum of requirements. These range from the regime's legitimization via the inculcation of a desired value system, appropriate adaptations in (or perversions of) traditional political culture, and coaptation of national symbolism, to the stimulation of desired behavioral traits and activities while discouraging prohibited ones. In the case of communist governments which were imposed by force of arms (some almost literally "parachuted" into their countries), this type of conditioning has had all the earmarks of warfare, though its scope is of necessity much broader and its nature different from those of other such governments which reflect majority political and social preferences and build on a shared value system by concentrating on defining and operationalizing desired tactical objectives. "Parachuted" communist regimes have had to contend not only against the traditional political and social system, as in the case of all other ruling Leninist parties, but were also in conflict with local nationalisms. An entirely new base had to be built in a hostile territory.

A comprehensive analysis of psychological warfare conducted by East European communist governments against their societies is well beyond the scope of this chapter (and capabilities of its author). What the chapter attempts is to treat the theme in just one country and in a rather eclectic fashion, selecting aspects that are most familiar and seem most relevant, appropriate, and illustrative.

No effort is made to discuss instrumentalities of psychological warfare. The decision mechanisms are not a matter of public record, while the implementation mechanisms, i.e. the broad spectrum of communist socialization agencies and their work methods, are well-known and have been described many times. Their scope extends over the sum total of social communications, and embraces all educational structures (from kindergarten to universities, inclusive of special networks of political education in civilian and military life), all party-directed social organizations (from youth organizations to trade unions and associations of creative intelligentsia), arts and literature, and written and electronic media.[1] One should include here also agencies of coercion for thought control and punishment of undesirable behavior. Together, and building on the destruction of all autonomous social organizations and institutions, these preclude access to alternative sources of information and conditions. But the regime still has to contend with agents of traditional socialization, such as the family and the church, and with the seepage of information from abroad, transmitted through direct contacts or by Western broadcasts.

The country chosen for the chapter is Poland, and the analysis of changing popular perceptions will focus on three major themes: *historical perspectives, attitudes towards the Polish armed forces,* and *attitudes toward the social system.*

Poland was chosen as a case study for several reasons, subjective and objective. Subjectively, it is the country best know to the author, including the language and history, through personal experience and research of many years. In objective terms, it is the only country in the

Soviet bloc with a comprehensive series of public opinion polls for the period between 1956 and 1988, and highly developed research in sociology and social psychology. The results of the polls and the research have been accessible, because, since 1956, Poland has been the most open of all the bloc countries as well.

Another reason for the choice was that among member states of the Soviet bloc, Poland has been among the most resistant to the absorption of the Marxist-Leninist system and values. Historically a product of Western cultural influence and, through centuries, the "Defender of the Faith" against Orthodox Russians, Moslem Turks, and Lutheran Swedes and Germans, the Poles integrated Roman Catholic religious values into their national identity. The establishment of communism in Poland was not a product of indigenous growth, but a result of occupation by the Soviet Army at the conclusion of World War II. The resistance to communism was thus strengthened by traditional hostility towards the Russians, made more bitter because of the Soviet treatment of the Poles throughout the war. The Soviet Union started by dividing the country in agreement with Nazi Germany under the secret protocols of the 1939 Ribbentrop-Molotov Pact, and ended with the physical destruction, in 1944-45, of the Polish underground state loyal to the Polish government in exile in London, comprising the political, administrative, and military structures which, by all accounts, were the strongest and most effective in occupied Europe.[2]

Post-1945 history of Poland indicates that the level of resistance both to the system and to the Soviet alliance has been uniquely high. Thus, while communist psychological warfare there has generally been a failure, examples of even a partial success in affecting the perceptions of the Polish people are significant, and quite a few such examples could be found in the 1950s, 1960s, and 1970s. Even these, however, proved to be short-lived in the light of political changes in the 1980s, beginning with the emergence of Solidarity and ending with the reforms made possible by Gorbachev's *perestroika*. Inevitably there were transformations in the value system and in popular perceptions engendered by the war and its resulting dislocations, and attendant on the imposition of a new political system and its drastic new economic and social policies. But few, if any, went in the direction expected and desired by the regime.

The cumulative results, nonetheless, were compared by one Polish sociologist, to a "grinding mill."[3] There was an unprecedented increase in social mobility, both horizontal and vertical. Millions of people were uprooted, shifted, and geographically relocated under both the Nazi and the Soviet occupations and in the 1944-45 settlement. Vertical social mobility also reached massive proportions. A large part of the traditional upper strata and the intelligentsia were destroyed because of repressions both by the Nazis and by the communists. A rapid upward movement of the lower social strata was stimulated by new social and economic policies, such as nationalization of industry and land reform, and preferential treatment given to workers and peasants on a class basis in access to jobs and education. By 1948 all autonomous social and political organizations were destroyed (with the exception of the Roman Catholic Church which was persecuted), and the imposition of monopoly on social communications was coupled with intensive indoctrination in Marxism-Leninism and the rewriting of national history.

HISTORICAL PERSPECTIVES

The whole of Polish historiography became subject to revision and reinterpretation. Special attention was paid to the most recent period, in order to undermine opposition and to build a modicum of legitimacy for the new regime. Key problems targeted in this context were the perceptions, shared by most Poles, that the new communist government was a creature of a foreign and hostile power (namely "the Russians"), that it was imposed by force of foreign (Soviet) arms in the face of a determined and prolonged resistance, and that it had little or no support in the country. The goal was to build up a new image of the communist government as a genuinely Polish government which represented all the "progressive" forces in the country, meaning that only "bad" elements, such as "fascists," "bandits," representatives of the old "reactionary" social order, and "agents" of imperialist powers, opposed the new order. The new paradigm included also an image of the Soviet Union as the friendly power, whose assistance and support had been essential for the restoration of Poland's sovereignty and remained essential to guarantee the nation's well-being, development and survival. The impression conveyed was that the Soviet Union was, in fact, Poland's best friend in the world.

The subject is obviously too broad and too complex to be reviewed in a short chapter. On the whole it is the author's perception that the more than 40 years of intensively conducted psychological warfare designed to create new attitudes have not been successful. Popular attitudes in the 1970s and 1980s clearly indicated the refusal of a majority of the Poles to identify with communism and the communist political system, and revealed that they saw themselves and the government in the context of a "we" and "they" dichotomy, which absolved them both of any loyalty and commitment to regime-sponsored institutions and policies and of any responsibility for the conduct of affairs, and justified a withdrawal from public life. Many people saw the period following 1945 as the "occupation," and perceived the Soviet Union as the main obstacle to the country's sovereignty and the main threat to its survival.

Nevertheless, popular perceptions did change, and the changes were the more important because they were incremental and thus barely noted. A few examples related to the origins of the system will suffice here, although many more can be found. Take, for example, the myth of Poland's "liberation" and the tacit acceptance of the new Polish-Soviet boundaries which left the whole of pre-1939 eastern Poland to the USSR.

Until the late 1980s not a single official publication—beginning with children's books and ending with research monographs—has deviated from the norm in describing the entry of the Soviet troops, and the accompanying Soviet-sponsored First Polish Army, into Polish territory in 1944 as the "liberation." To a Western reader, the image immediately evokes the liberation of France or the Low Countries by the Allied armies, and a similar general image has been mechanically fostered—via endless thoughtless repetitions—in the minds of the generations of Poles who were not among those directly affected. The "liberation" has been portrayed, moreover, as both a "national" and a "social" liberation, the first from a German yoke, the second from the yoke of domestic class oppression. Technically the image is correct, because the Soviet armies traversed Poland in a victorious campaign against the retreating armies of the Third Reich, and thus relieved the country from the German occupation. The picture of social liberation also seemed convincing to sons and daughters of the workers and peasants as they were invited to help in the construction of a new "socialist" Poland.

What got lost in the shuffle, however, was the liberation's by-product, namely the destruction by the Soviet Union of the whole fabric of independent Polish political life, including the annihilation of the 400,000-strong underground forces, the Home Army. The Home Army stood ready to rise behind the German lines within the framework of an operation code-named "Tempest," adopted by the leaders of the Polish underground state in consultation with their London-based superiors. The Tempest was a desperate attempt to safeguard Poland's sovereignty in the face of Soviet determination to impose a post-war settlement suited to Soviet interests which, at this point of the war, was acquiesced to by the Western Allies. As the front approached prewar Polish boundaries, Home Army units were instructed to intensify anti-German sabotage, to mobilize for an open uprising in the German rear as the front moved in, and to coordinate military action with Soviet units which were to be treated as allies and fought only in self-defense. At the same time Polish civilian authorities were to emerge and take over the administration of the liberated areas. Tempest went into effect as Soviet troops crossed the 1939 Polish-Soviet boundary in January 1944, first in the Volhynia, Vilnius, and Lwow regions, then farther west.

Militarily the operation was a success, but politically it turned into a disaster for the Poles, as the military and political infrastructure of the government which emerged from the underground was systematically destroyed by the Soviets. Frontline Soviet troops readily accepted Home Army's cooperation, but they were followed by NKVD troops who arrested Home Army commanders and leaders of civilian Polish authorities; the soldiers were either dispersed, forcibly incorporated into the First Polish Army and/or penal battalions, or were sent to the gulag. As Soviet actions became known, many Home Army units in central and western Poland retreated into the forest. The Warsaw Uprising of August 1, 1944 (when Soviet forces reached the right bank of the Vistula across from the city), was the final desperate effort—counting on Western support—to establish Poland's sovereignty as Germany went down to defeat. In response, Soviet troops were ordered to halt their advance, and watched while the Germans pulverized the city, the desperate resistance of which lasted 63 days.[4]

The existence of the Home Army has never been denied in Polish communist sources, but its actions have been both minimized and tarred with a brush of "counterrevolution."[5] The communist Polish underground, the Peoples' Army, on the other hand (the size of which is variously estimated at between 10,000 and 60,000 maximum, even by official sources[6]), has been credited with most of the heroic anti-German exploits, along with numerous Soviet "partisan" units, which operated on Polish territories. In reality, the "partisans" were Soviet diversion and sabotage specialists dropped behind enemy lines. Soviet sources estimated that some 10,000 Soviet citizens served in Polish communist partisan units, and that there was a total of 12,000 Soviet "partisan units" in Poland.[7] This is an astounding number, considering the People's Army total; thus, it is no surprise that these figures are rarely if ever cited in Polish sources. Constant repetition did achieve results and a perception that Poland was indeed "liberated" by the Soviet army gained wide currency. The irony is that the acceptance of the "liberation" myth seems to have benefited the image of the Polish People's Army, the LWP (created by a merger, in July, 1944, of the First Polish Army with the partisan Peoples' Army), which was a part of the Soviet advance, rather than that of the Soviet Armed Forces.

The acceptance of Polish-Soviet boundaries imposed by the Soviet Union—the legal basis for which is the Nazi-Soviet Pact of 1939 and a phony referendum conducted after the area was incorporated into the Soviet Union—was one of the crucial obstacles to the rapprochement between the Polish government in exile and the Soviet government after Nazi Germany invaded the Soviet Union in June 1941. (The other was the demand for information on the almost 15,000 Polish officers taken prisoner by the Soviets, who simply disappeared. The bodies of 4250 of them were found by the Germans in 1943 in mass graves in the Katyn forest near Smolensk, at which point the Soviet Union announced that they were killed by the Germans.[8])

The government-in-exile refused to compromise on the 1939 boundaries until the end, as reflected in the sphere of operations of the Home Army noted earlier. But the Soviet (and thus also the communist Polish) view has been that the Ukrainians and Belorussians in eastern parts of Poland were saved from the Germans and restored to their national republics by the Red Army's entry in 1939. It was only in the 1970s and 1980s that Polish sources began to discuss the "excesses" committed by Soviet authorities against the Poles in these areas—such as the wholesale deportations to the Soviet Union—but these are blamed on Stalin.[9] At the same time the myth of "liberation" has also legitimized the new boundaries.

The formation of the Polish Committee of National Liberation, the PKWN (which was the nucleus of the future communist Polish government) was planned to coincide with the entry of the Soviet troops and the Polish First Army into the "Polish territory," which, according to the Soviet and communist Polish interpretation, happened in July 1944, when the river Bug was crossed, and Lublin, the "first large Polish city," was "liberated." It was in Lublin that the PKWN was first unveiled. The date of the unveiling, July 22, 1944, became the new Polish national holiday, and the committee was immediately recognized by the Soviet Government as "the true representative of the Polish people." In fact the PKWN was formed in Moscow and flown to Lublin and, as pointed out by a British historian, was fully a creature of Moscow, composed half of "NKVD appointees" and half of "Soviet employees."[10] But the legend has been built subsequently from kindergarten on up, and the holiday has been celebrated now for 44 years with few questions asked.

The importance attached to the symbolic value of the "Polish Army" first entering the "Polish territory" is explicitly admitted in Soviet and Polish military sources that discuss the "political mission" of the First Polish Army. General Bordzilowski, one of the Soviet officers of Polish origin seconded to the First Army and a chief of the Polish General Staff from 1949 to 1965 (when he was replaced by General Jaruzelski and retired back to the Soviet Motherland) recalls that the Soviet command "saved the First Army (from combat) in order . . . to move it into the forefront of the fighting once the Polish border was crossed. . . . The key function was to allow the heroic but martyred city of Warsaw . . . to be liberated with the participation of the First Army, the new Army of a new Peoples' Poland."[11]

The mission is confirmed by General Shtemenko, wartime Soviet chief of staff: "The Soviet Command has been sparing (the First Polish Army) and only now (in the Warsaw region) decided that it had the right to bring it into action. The Polish comrades were given a chance to demonstrate their patriotic feelings directly under the walls of Warsaw. . . ."[12] The gesture might have been more appreciated by the Poles had it come before October 5, 1944, when the Home Army in Warsaw capitulated to the Germans. First Army's heroism in January 1945 was wasted on empty ruins of Poland's capital.

Another myth that gained currency by repetition is the myth of a "civil war," which, according to communist historiography, raged in Poland in 1944-47 between the (good) forces of the new government on the one side, and the "baddies"—The Home Army and its successor organizations, "NIE" (Independence) and "WIN" (Freedom and Independence), the NSZ (right-wing National Armed Forces that had refused to affiliate with the Home Army), and the Ukrainian revolutionary UPA, on the other. It is estimated that some 80,000 armed partisans operated in Poland in the first half of 1945,[13] the peak year of the struggle. But on the communist side, most of the forces involved in the pacification were not the Polish units but Soviet NKVD troops, which freely operated in the country. The NKVD was aided by the still operating Soviet "partisans." Polish communist participation in the wiping out of the remnants of national Polish and Ukrainian resistance was limited to newly formed security troops, and a few select military units. (General Jaruzelski participated in the pacification of the Ukraine as a young officer.) Clearly, the struggle was not a "civil war" but the wiping out of national resistance by foreign troops.

Many resistance sympathizers were to be found in the ranks of the LWP, especially in the Second Army, which was formed on the basis of draft with many of its officers recruited from among People's Army and some Home Army veterans. The conversion of the latter was initially welcomed; the Home Army was officially dissolved January 19, 1945. The rate of military desertions "to the forest" can be judged by the numbers amnestied in 1945 and 1947: LWP deserters numbered 7448 men.[14] Polish communist leaders freely discussed the problem of mass desertions, and in fact had to defend themselves against Stalin's accusations that the LWP was unreliable, and that the Poles were "ungrateful" and treated Russians in Poland "badly."[15]

There is ample evidence that Soviet forces acted on their own authority in dealing with "hostile elements" on Polish territory, although a legal fiction was observed when a formal treaty was signed between the USSR and the PKWN on July 26, 1944, giving Soviet forces full control over civilian security in the Soviet rear. Soviet sources reveal that it was the Stavka (the Soviet High Command) which ordered Soviet commanders in Poland to destroy armed resistance, (to make possible "the normal activity of the PKWN" and to allow "for the mobilization of youth" for the Polish army) and to conduct all other operations that were necessary, such as appointment of local officials, confiscation of foodstuffs, and so on. Even the famous arrest of the 16 leaders of the Polish underground in March 1945 which resulted in their trial and imprisonment in Moscow, was apparently carried out by the Soviet armed forces without even bothering to inform the provisional Polish government (a new reincarnation of the PKWN).[16]

Polish national memory has retained the negative picture of the Soviets, despite all efforts to the contrary (as shall be seen below), but their role in the suppression of mass national resistance, which is what the "civil war" really was, seems to have disappeared under the new myth. It gained currency through repetition, and thrived on many personalized and localized stories. In the process, the numbers of, and the atrocities committed by, "fascists" and "counter-revolutionary bandits," miraculously multiplied, while "true patriots" were praised for their heroism and sufferings.

None of these myths, now seemingly incorporated into the Polish psyche, can be found on the list of the "white spots" *(biale plamy)*, which have come to the forefront of the revision of the historiography of Polish-Russian relations ushered in since Mikhail Gorbachev became the first secretary of the CPSU. Instead, attention has centered on the events of which the reinterpretation was not successful, such as the 1944 Warsaw Uprising, and the Katyn Massacre.

In the case of the Warsaw Uprising, revisionism started in the 1970s. Because it proved impossible to obliterate heroic memories of Home Army's struggle, an effort has been made to coopt the heroism while continuing to condemn the leadership. The fighters of the Home Army (the People's Guards' participation in the uprising has also been highlighted), have been praised for their patriotism and bravery, and have been portrayed as the victims of their treacherous and criminal leaders. General Jaruzelski himself has participated in commemorations of the uprising's anniversary, and the communist government even created a medal to decorate the surviving fighters, few of whom, however, were willing to accept the award. As for the Soviet Army's inaction, it is still blamed on exhaustion and overextended lines of communications.

In the case of the Katyn Massacre, for years the Soviet excuse (repeated by Polish communists) has been to blame the killings on the Germans, despite powerful evidence to the contrary. The deception was carried to the length of building a monument in Warsaw to the martyred officers, with an inscription once again ascribing the murder to the Germans. But the clamor for the admission of Soviet guilt, attendant on the winds of change that have recently swept the Soviet bloc, found supporters even among Polish party historians. Finally, a spokesman for the Polish government officially admitted that the officers were, in fact, murdered by the NKVD. Polish members of the official Soviet-Polish Commission of Party Historians for the Study of Relations Between the Two Countries published their official findings on the subject with the same conclusion.[17] The admission was made possible because of Soviet glasnost, and new opportunity to blame everything on Stalin and Beria (the head of the NKVD). But, as of the date of this writing, Poles were still waiting for an official Soviet confirmation of Soviet guilt, or for any information on the fate of the Polish officers who also disappeared but were not found in Katyn. Soviet authorities, nevertheless, permitted the Poles in Smolensk to build a religious memorial at the scene of the mass graves in Katyn; the inscriptions, predictably, blame the Germans.[18]

"Our Boys"—A Success Story?[19]

Polish communist military historiography has always portrayed the Polish People's Army, the LWP, as the direct heir of the Polish armed forces of the Second Polish Republic (1918-39), the Home Army, and the Polish Forces in the West, as well as their historical predecessors. This image seems to have "taken," and is thus a success story of psychological warfare.

In fact, as noted above, the LWP originated in a merger between the Polish units formed in the USSR and a handful of communist partisans, both heavily infused with Soviet citizens. In organization, training, and equipment the LWQ has been a clone of the Soviet armed forces. In July 1944 it numbered 107,000 men (as compared to the 400,000 Home Army); its numbers were subsequently quadrupled by instituting the draft, and reached 400,000 in May 1945 when it participated in the final Soviet assault on Berlin. It may be worth noting that LWP's current strength is exactly comparable: it had 406,000 men in 1989,[20] reflecting steady expansion since the low demobilization point of 100,000 men in 1948, although cuts were envisaged under Warsaw Pact reform plans for the 1990s.

When still in the USSR, about two-thirds of the First Army's professional cadre were composed of Soviet officers (few professional Polish officers survived Soviet captivity, and in any case most were politically unreliable). In absolute numbers, some 20,000 Soviet officers served in the First Army during the war, 36 of whom were general officers. According to General Bordzilowski, who was one of them, "they were not seconded as advisers but took over command positions directly"[21]; many were ethnic Poles. By 1945 their proportion in the LWP was reduced to one-half of the total and, by 1948, to an approximate one-third of the command cadre, as the ranks were filled partly by training young Polish officers and partly by admitting some Home Army officers and professional officers returning from the Polish Army in the West. But, beginning in 1949 when Soviet Marshal Konstantin Rokossovski became the Polish minister of Defense, the "unreliable" elements among the professional cadre were savagely purged as a part of a new policy of modernization. These included many officers of the wartime People's Army. Estimates for the 1949-56 period indicate a saturation of the LWP command, staff, and political cadre with a new influx of Soviet officers for an approximate 90 percent of the total.

The upheaval of 1956 was marked by an effort at the emancipation of the LWP from under the Soviet tutelage, and a restoration of the force's national character, under Marian Spychalski a new Defense minister and a protegée of the new Polish leader, Wladyslaw Gomulka. The first was short-lived, but the polonization of the force remained a permanent feature. Most Soviet officers returned home, although some remained in Poland permanently. General Bordzilowski, who was appointed the LWP chief of staff by Rokossowski and held the position until the mid-1960s, is a case in point, and there were others.

A new period of modernization of the Polish forces, and their closer integration into a unified system of command of the Warsaw Pact, began in 1960, when Marshal Grechko came in as the Pact's commander-in-chief. New mechanisms were developed to ensure operational integration of East European armies into a regional military system under Soviet command. These included the common (Soviet) military doctrine and the language of command (Russian), a joint command system (with key positions held by Russian general officers), common structural and operational features, secondment of best-trained, best-equipped units from the national to Joint Warsaw Pact operational command, a system of selection and training of professional military cadre, and an integrated political education system, all of which were derived from Soviet experience and subject to control by Soviet high command. Advancement in the command and political stream and into general officer ranks for the East European cadre, for example, has been contingent on training in Soviet staff schools. Thus, the LWP has become a part of a military system that transcends the national Polish state.

Nonetheless, the national image of the Polish soldiers has been of major importance to the regime. The "patriotic" role of the First Army was deliberately planned and implemented by the Soviet high command, as noted earlier. But the Polish population remained unconvinced, despite the adoption of traditional trappings and national symbols, which include a Roman Catholic chaplaincy and a practice of field mass dutifully attended by communists and Soviet officers in Polish uniforms. In the immediate post-1945 period there was much public derision of the "new" army and their "new" officers, and LWP prestige sunk even lower in the Rokossowski period.

But the prestige and national character of the army were both restored in 1956 and gradually gained ground in the 1960s—with temporary setbacks caused by LWP participation in the invasion of Czechoslovakia in 1968, and by the use of troops in suppressing the Baltic strikes in 1970—and in the 1970s. In the 1970s, in particular, a professional military career was made more attractive. LWP's favorable image was also strengthened by an intensive manipulation of patriotic symbols, and by appeals to historical martial traditions. One perceptive observer noted the proliferation of traditional martial symbolism in the flood of historical films and publications and reached the conclusion that all were being manipulated for profoundly non-Polish purposes.[22]

The effectiveness of patriotic propaganda on the military front became visible in the Solidarity period (August 1980-December 1981) when both General Jaruzelski and the LWP proved to be immensely popular. Jaruzelski's popularity in particular was based on a widely held image of a patriotic, apolitical, and austere professional soldier, who allegedly told the rest of the Politburo in August 1980 that he would not use Polish soldiers to shoot at the Polish workers. Solidarity membership assumed that the army would naturally defend Polish interests against outside intervention, and published posters telling the "General" that "the Nation and the Fatherland are waiting," and appealing to him to form a joint "front of social defense . . . against lawlessness." The Solidarity newspaper described the LWP as "our sons" who "are of us, for us and ours, because they are ready to defend us."[23] One of the ironies of the period was that when martial law was imposed in December 1981, the justification for it was phrased precisely in terms of "defending" Poland against a worse fate (namely a Soviet invasion). The kernel of truth contained in the assertion made it plausible to at least part of the population, and also served to refurbish Jaruzelski's image as a patriot.

Both official and unofficial opinion polls conducted in the Solidarity period showed a high level of trust in the army, as contrasted with a low level of trust in the party. A May 1981 poll conducted among the public at large listed 89 percent of respondents expressing trust in the LWP (as compared with 32 percent trusting the party). The LWP also scored high with Solidarity members. In an October 1981 poll, 69 percent of the members questioned expressed trust in the armed forces, as compared with only 7 percent for the party. In both cases the only two institutions which scored higher in both polls were the Church and Solidarity.[24] A survey conducted by the Public Opinion Research Center of Polish Radio and Television in November 1981 tested attitudes towards the military task groups deployed in the countryside ostensibly to curtail abuses, but in fact as a move preparatory to the imposition of the martial law. It showed a top approval rating of 91 percent.[25]

The imposition of martial law by the very army that was supposed to defend Solidarity was a profound shock to the population. Polling was difficult after martial law, and the results of the ones taken were not immediately published. Had the results been favorable to the army and to the new Council of National Salvation (WRON) led by General Jaruzelski, it is fair to assume they would have been made public. The director of the official polling agency announced in December 1982 that a new poll showed the military "maintaining very high authority,"[26] but a secret poll conducted by *Paris Match* with 600 respondents in May 1983 showed a dramatic decline in the regard for the armed forces. Only 2 percent of respondents agreed that the forces "best represented" Poland, and only 6 percent thought that General Jaruzelski was Poland's "last hope."

But Solidarity (with 14 percent), and the Church (with 21 percent), did not score all that well in the representation question either, and Walesa himself received only 6 percent on the "last hope" question, as compared with 28 percent for the Pope, 27 percent for the Virgin Mary, and 7 percent for President Reagan.[27] By the mid-1980s there was some improvement in the LWP and Jaruzelski's image, but nothing to compare with their premartial law standing: only about one-fourth of samples of public opinion surveys assessed the LWP as an institution standing above party quarrels and expressed confidence in General Jaruzelski.[28]

The December 1981 imposition of martial law was dressed up in patriotic rhetoric from the beginning to the end. In telling the country that he had imposed martial law in a six A.M. speech December 13, 1981, General Jaruzelski used high-flown phrases, ending with two lines from the Polish anthem: "Poland will not perish while we live," the implications of which were that "we, the military, will protect Poland!" Subsequently, the official media, the military press in particular, castigated Solidarity for vainly abusing both the national symbols and national interests. The action itself was carried out primarily by special units of militia (ZOMO) and the security troops, with the military mostly kept in the background. Thus, the main wrath of the population turned against the former, and people were reported saying that "the soldiers did not really want to do it," or that they "could not help it."

The furious revival of interest in Polish national history in the Solidarity period, coupled with a virtual breakdown of the regime's monopoly of communications, seriously undermined the effects of the regime's patriotism campaign and the WRON's response has been to mount an intensive effort to coopt the Polish past, particularly its martial traditions. Honor guard details put on traditional diamond-shaped hats, and cadets in officers schools put on historical uniforms at ceremonial occasions, the uniforms worn by the troops of the 1815 Congress Kingdom who were involved in one of Poland's major risings against the Russians, the November Revolution of 1831. In a ludicrous exercise members of the Sejm (the national assembly) debated whether or not a crown (removed in 1944) should again be placed on the head of the eagle, the national symbol, so that it could not be mocked as a crow (WRON, the acronym for Jaruzelski's National Salvation Committee means a crow in Polish). The campaign involved setting up historical monuments and celebrating popular national anniversaries such as that of the 1944 Warsaw Uprising.

The campaign to coopt Polish historical traditions had no choice but to coopt also their anti-Russian content, but this problem was bypassed by directing its thrust against czarist Russia, rather than the Soviet Union. In the indoctrination of the armed forces, the memories of Polish anti-Russian struggles are recognized and have been openly exploited: Czarist Russia was the enemy because it exported both national and social oppression. Russian and Polish working people, on the other hand, have always been friends, and it is only thanks to the Great Socialist October Revolution and the establishment of the Soviet Union that it became possible for Poland to build its new socialist society.

"Our Soviet Friend"—A Failure?

None of the benefits of LWP's growing legitimacy seem to have rubbed off on its Soviet friend, but not for lack of trying. For years, from the cradle to the grave, practically every Pole was subjected to stories extolling "our Soviet friend." These varied in sophistication from kindergarten poems to learned treatises, praising the greatness of the Soviet Union, expressing gratitude to Russian workers who, by being the first to build a socialist society, extended its benefits to the Poles, describing Soviet friendship and selfless assistance and the benefits that have accrued from Poland's membership in the "socialist community." But all the pro-Soviet lore had little impact on anti-Soviet attitudes. Their existence was openly deplored in official sources in the 1970s, for example, and a campaign condemning hostility toward the USSR was mounted by the government in the Solidarity period, warning specifically of damages such attitudes might bring to Polish-Soviet relations. Much of the official fervor had to do with fostering the perception, never directly articulated but nevertheless clearly understood, that any threat to "socialism" in Poland would bring on Soviet intervention.

In this the regime has clearly succeeded. In the official polls of 1980 and 1981, 49.6 and 36 percent of respondents, respectively, identified the USSR as the greatest threat to Poland. Other socialist countries were seen as the next highest threat by a respective 13 percent and 20.2 percent of each sample.[29] This meant that over one-half of all respondents in both polls identified the country's official allies as, in fact, its greatest enemies. Post-martial law informal polls expressed similar sentiments. A survey of Solidarity members in the Warsaw region carried out early in 1982, showed 54 percent of the respondents agreeing that the military coup took place because of Soviet pressure,[30] and in the *Paris Match* poll, the USSR and two other Warsaw Pact 'allies" led the list of Poland's perceived "worst enemies": USSR with 29 percent, the GDR with 24 percent, and Czechoslovakia with 18 percent of respondents. Thus, as in earlier polls, 70 percent of the respondents saw Poland's enemies within the "fraternal alliance," and 60 percent felt that Soviet intervention was still a possibility.

Behind the facade of friendship rituals, similar perceptions seem to have motivated the country's leaders. There was a certain note of urgency in General Jaruzelski's argument at the June 1981 Plenum of the Central Committee of the Polish United Workers Party (PUWP) in the discussion of the Soviet letter of June 8 (believed to have been an ultimatum) emphasizing the crucial importance of maintaining stability in Poland:

> Our position in the Warsaw Pact defines a very specific degree of national and international interdependence. Thus, what is happening here, the destabilization of Poland, threatens to upset the balance in a context far broader (than the local) one.

Further, he pointed out that the emphasis in the Soviet letter on Poland's boundaries and independence being "guaranteed" for a "concrete, socialist, and friendly Poland" was "not coincidental," and those who did not see it were blind.[31]

Clearly, from the point of view of the survival of the regime, the important message to convey to the people was the perception of *limits* imposed on popular (and the regime's) behavior by "Big Brother's" looming shadow. Whether or not one also "loved" the Soviet Union was, in the final analysis, irrelevant. The whole weight of mass persuasion directed at extolling Soviet virtues have done little to change Polish collective thinking about the eastern neighbor, but the perception of the limits that Soviet might and presence and the Soviet alliance has imposed on Polish sovereignty was very well absorbed indeed, and thus should be counted as perhaps the one unambiguous success of psychological warfare. Although the direct message itself was rejected, its implied meaning was well integrated and in fact became the guiding factor in Polish behavior, on both sides of the political divide.

It was very much present in the minds of the Polish communist leadership in the 1980-81 upheaval. It was also extremely well rooted in the perceptions of the leaders of the Polish opposition. Thus, it was no coincidence that Solidarity activists described their movement as the "self-limiting revolution," and that the phrase, Poland's geopolitical position," has become the code word for explaining why, *rebus sic stantibus,* the country has to remain a member of the Warsaw Pact, and why, regardless of the intensity of internal pressures for systemic change, some semblance of "socialism," as defined in the final analysis by Moscow, had to be preserved in Soviet-dominated Eastern Europe.

The perception of the limits, and the crucial requirement of avoiding destabilization, has been as valid in Gorbachev's era, as it was in 1981. But the perception of where the limits really are has become increasingly blurred, with a message from Moscow urging democratization and reform, albeit still within the framework of "renovating socialism." Thus Solidarity revived and "reformers" emerged at the cutting edge of the Polish party, leading to a national agreement on reform reached at roundtable discussions between government and Solidarity leaders. But the semi-free, multi-candidate elections of the summer of 1989 that resulted, carefully designed to preserve the party's leading role in the national assembly, backfired. Both the party and the opposition were caught unprepared by the wrath of the people who rejected all party candidates who ran for the elective seats, thus thrusting the opposition into a majority role, and giving it a capability to form a government.

Yet Solidarity leaders hesitated, fearful of crossing the invisible boundary. When they finally agreed to form a coalition government, they also made sure that General Jaruzelski did not lose the election, in the assembly, to the new constitutionally powerful presidency, which carries with it the chairmanship of the Council of National Defense. The general won by one vote, but only because key opposition figures abstained. To both sides General Jaruzelski's presidency provides a guarantee of security. Moscow is assured that Poland (strategically the vital corridor) will stay in the Warsaw Pact, and Solidarity-led government feels protected against a possibility of an invasion. The perception of the limits survives still, even at these crucial political crossroads.

The Social System

Measured first in the 1950s and confirmed by subsequent polls, the values of Polish society showed two remarkable changes in the wake of the "grinding mill" of wartime dislocations and systemic pressures. First, there was a homogenization of values and attitudes into a single unified system, with no subsystems and no confrontations over worldviews. Second, the content of the value system indicated selective internalization of such new values promoted by the regime which were compatible with traditional perceptions, a combination that began to exert pressure on the national level by the mid-1950s.

These new values could be summarized as "socialism" but socialism of a different kind than orthodox Marxism-Leninism. They included the standard support for nationalization of industry and land reform (but not for collectivization), and for economic planning. But at the same time there was a strong emphasis on democratization and democratic values, such as respect for civil rights and imposition of legal limits on coercion, freedom of expression and political participation, and the right of workers to self-management—all of these in strong association with traditional religious and national values. High value was placed on egalitarian values. Egalitarianism, derived both from communist official theory and from historical traditions, proved to be an exceptionally strong value, both in terms of equality of opportunity and in terms of condemning inequalities in social distribution and class stratification. Equally strong support was given to a related value, that of social justice. The whole package was designated a "Polish road to socialism."[32]

General effects of the change did little to help the regime's legitimization, especially because of the growing gap between the official myth espoused by the PUWP and the reality as seen by the people. The contrast was perceived to be particularly glaring in the context of growing disparities of income and wealth. Conspicuous consumption by the privileged few, notably the members of the party's nomenclature, was particularly resented. There was also mounting resentment of political repressions and economic mismanagement, of recurring shortages of consumer goods, and a catastrophic housing situation. Alienation grew, characterized by reprivatization of life pursuits and indifference toward public life and communist institutions, and open cynicism; nevertheless opinion surveys confirmed that the values compatible with the "Polish road to socialism" remained in top position.

Surveys in six towns in 1966 and in 1977 showed the principle of egalitarianism moving from the sixth to the third place in the hierarchy of eight values; the first two, in both cases were the desire to keep the leaders honest and truthful and to avoid conflict.[33] In a March 1980 survey, equality ranked second, behind family, as the top value, and a September 1980 official poll revealed that equality and justice were the two most important social values in the country. A national survey of 1980 confirmed these findings, reflecting perceptions of ever greater inequalities. As one analyst noted,

> In this case, as with most of the values cherished by Poles, people generally favored the official regime values, but believed the reality of Polish society did not match the ideology and rhetoric. On egalitarianism, in particular, people wanted the rich and powerful to play by the rules that they, themselves, had propagated.[34]

Accumulated frustrations resulted in the explosion of 1980, but no improvement was possible in the tug of war between the movement and the government in the Solidarity period, and none in the "normalization" period either, after the imposition of the martial law, despite repeated efforts at half-hearted reforms. Opinion polls available confirm the high standing assigned to socialist values, while at the same time indicating an increase in alienation, a sense of hopelessness and deprivation, and a total disillusionment with the political system.

Findings by Polish social psychologists indicate that the open conflict between the regime and the people in the Solidarity period (1980-81) led to an appearance, in public consciousness, of a clear distinction between those who run the country, called simply "they," and all the rest, second class citizens, or "we," and introduced a new dimension of inequality, namely an inequality of participation in public life. The feelings of deprivation were not only extended to, but were also most strongly felt in, the sphere of public life, because more than 70 percent of the people felt that they could not realize there the values important to them. Political institutions created to carry out political, economic, and cultural changes basically failed in the performance of their tasks and thus also failed to satisfy either social or political needs of the population. This, in turn, led to an emergence of compensatory mechanisms: pressure for democratization at the grass roots level, opportunism, and a network of small groups and informal ties developed as alternatives to public institutions for the satisfaction of social needs outside formal political structures. The latter mushroomed after 1981.[35]

More recent polls revealed a catastrophic decline in popular confidence in the system in 1987-88. The number of respondents who felt the system was OK fell from 8 percent in March 1987 to 3.6 percent in December 1988, and those who felt that only some changes were needed, from 45.7 to 27.6 percent. On the other hand, the number of those who felt that only major changes would do increased from 35 to

43.1 percent, while those who rejected the system completely increased from 7.5 to 19.1 percent.[36] This was the background for the debates of the Xth Plenum of the Central Committee of the Polish party which met in December 1988, and in January 1989, the meetings which resulted in the consolidation of a new reformist leadership and proposals for electoral reforms along with the recognition of the right of the opposition to political participation, but still within the framework of the leading role of the Communist party. As noted earlier, the results of the election have caused this last bulwark of party rule to crumble.

It seems that the impact of the absorption of declaratory values promoted by a system that does not practice what it preaches has turned out to be highly dysfunctional to its legitimization. The effect has been opposite to that intended: It contributed to the loss of power by the ruling party, rather than to its maintenance. This outcome, of course, was conditioned by many factors other than the failure of the regime's psychological conditioning. Two of the major ones have been the failure of the system to perform and the green light given to reform by the Soviet Union.

CONCLUSION

It seems that the effect of psychological warfare directed at its own population by the Polish communist government since 1945 has on the whole been a resounding failure and in many aspects counterproductive. The grafting of desired ideas and stimulation of desirable behavior was successful only in selected cases and, at best, partially.

Conditions which contributed to success were: first, a predisposition in the traditional political culture to accept a given value (as in the case of egalitarianism); or, second, promotion of a manifestly true character of perceptions (as in the case of the limits imposed by the Soviet alliance). Good results could also be achieved by building on or adding to a concept that was perceived to be a true one (such as the patriotism of the military), in which case the desired effect could be achieved by the manipulation of the concept.

Success was impossible, on the other hand, if a value or behavior promoted was totally alien to Polish political culture or challenged any of its core values, such as religion or nationalism, or was in conflict with deeply ingrained perceptions, such as Polish Russophobia.

In any event, as the events in Poland in the 1980s convincingly demonstrate, changes in the collective psyche proved reversible as conditions changed, and the regime's monopoly of communications was breached by alternative sources of information and influence, domestic and foreign, while its general effectiveness declined. Also, internalization of values that are declared to be valid but are in fact honored in the breech proved highly dysfunctional to the main goal of the whole exercise, that is, system maintenance.

In the final analysis, a strategic warfare success in psychological conditioning of a society would require total isolation from any alternative sources of influence, a condition which proved impossible to achieve even in Stalin's Russia, where traditional agents of socialization managed to survive and function, and in fact became increasingly important as the mobilizing power of the Marxist-Leninist myth declined, and the system stagnated.

* I am grateful to two of my colleagues, Yosef Lapid of Carleton University, and Melvin Croan of University of Wisconsin–Madison, for their constructive comments. Needless to say they bear no responsibility for the final product.

NOTES

1. For a comprehensive recent treatment, see F.C. Barghoorn and T.F. Remington, *Politics in the USSR*, 3rd ed. (Boston: Little Brown, & Co., 1986).

2. See Stefan Korbonski, *The Polish Underground State: A Guide to the Underground, 1939-1945* (New York: Columbia University Press, 1978), and T. Bor Komorowski, *The Secret Army* (New York: Macmillan, 1951).

3. Stefan Nowak, "Values and Attitudes of the Polish People," *Scientific American* 245 (July 1981): 45-53.

4. See Bor-Komorowski, op.cit. The commanding general of the First Polish Army (Z. Berling) sent a few units to Warsaw across the Vistula on his own authority. He was relieved from command.

5. It was also accused of inactivity. The treatment is reminiscent of that meted out by the Yugoslav Partisans propaganda machine to Drazha Mihailovic's Chetniks, although in the scale of activities and organization the Chetniks did not even come close to the Home Army. For a different point of view see Nora Beloff, *Tito's Flawed Legacy* (London: Gollancz, 1985).

6. M. Jurek and M. Skrzypkowski, *Tarcza Pokoju: Ixlecie Ukladu Warsza Wskiego* (Warsaw: MON, 1975), 309.

7. I. Semiriage, A.V. Antosiak, and P.M. Derevianko, *Zarozhdenie Narodnykh Stranuchastnits Varshavskogo Dogovora, 1941-1945* (Moscow: Nauka, 1975), 103, 105.

8. A communiqúe of the Special Soviet Investigation Commission of January 24, 1944 (so-called Burdenko Commission). This remained the only official Soviet pronouncement on the subject as of October 1989, although in 1989 more than one official Soviet source hinted at Soviet guilt.

9. For details see Jan T. Gross, *Revolution from Abroad; The Soviet Conquest of Poland's Western Ukraine and Western Belorussia* (Princeton: Princeton University Press, 1988).

10. Norman Davies, *God's Playground: A History of Poland*, Vol. II (1795 to the Present) (New York: Columbia University Press, 1982), 558.

11. Jerzy Bordzilowski, "Rola Armil Radzieckiej w tworzeniu Ludowego Wojska Poiskiego i w wyzwoleniu Polski," *Wojskowy Przeglad Historyczny*, no. 3. (1978): 38 and 40.

12. S. Shtemenko, *Generalnyi Shtab V Gody Voiny* (Moscow: Voenizdat, 1968).

13. A. Polonsky and B. Drukier, *The Beginnings of Communist Rule in Poland* (London: Routledge and Kegan Paul, 1980), 106.

14. Cited in Korbonski, 245.

15. Polonsky and Drukier, 356, 361.

16. Davies, 471; I. Semiriaga, et al. 120-21; and Polonsky and Drukier.

17. *Polityka*, 19 August 1989.

18. Information from a visitor from Poland who actually visited the memorial and made photographs.

19. The discussion in this section is largely based on the documentation and analysis from T. Rakowska-Harmstone, "Poland," 1-271, in T. Rakowska-Harmstone, Christopher D. Jones, and Ivan Sylvain, *Warsaw Pact: The Question of Cohesion*, Phase II, Vol. 2, *Poland, German Democratic Republic and Romania*, ORAE Extra-Mural Paper, no. 33 (Ottawa: National Defence, November 1984).

20. *The Military Balance 1988-1989*. (London: IISS, 1989), 50.

21. Bordzilowski, 37.

22. Kazimierz Brandys, "Nierzeczywistosc," *Zapix I* (London: Index on Censorship, first Polish edition, January 1977), 33-34. A publication circulating outside censorship.

23. *Solidarnosc*, no. 5 (1981): 3.

24. Cited in David S. Mason, *Public Opinion and Political Change in Poland, 1980-1982* (Cambridge, MA: Cambridge University Press, 1985), 118.

25. *Polityka*, 12 December 1981, 2.

26. Mason, 211.

27. Biuletyn Informacyjny, no. 77, September 1983.

28. Maciej Nowiecki, "W cieniu na wulkanie" (In the Volcano's Shadow), *Aneks*, 49/1988, 59-79.

29. Mason, 177.

30. Mason, 214.

31. *No We Drogi*, no. 7, 1981.

32. Based on Nowak.

33. Mason, 63.

34. Mason, 63-65.

35. Jadwiga Korelewicz and Edmund Wnuk Lipinski, "Visions of Society, Differentiation and Inequalities in the Collective Consciousness," *Sisyphus*, Vol. V (1988).

36. *Polityka*, 21 January 1989, 6.

8

The French Army and Psychological Warfare

Pierre Lacoste

The long-term view of psychological operations and political warfare can be presented in four main points. First, psychological warfare has had tremendous impact on the French Army. Second, psychological warfare is just another branch of strategy and politics. Third, psychological warfare is as old as war itself and is affected by history and culture. Fourth and finally, different actors use a great variety of psychological tools, methods, and courses of action. Psychological warfare in the nuclear age, especially given the new prospects for radical changes in international relations, is a subject for serious concern.

PSYCHOLOGICAL WARFARE AND THE ARMY

The experiences of the French Army present numerous examples of the impact of psychological warfare, which has been of significant importance to the French Army since World War II. The German occupation of France after the military defeat of June 1940 and the armistice resulted in a mixture of collaboration and resistance. The situation created many psychological problems among the military community; the main problem being whether to obey the legal Vichy government, which was more and more controlled by the Germans, or to join DeGaulle, who had been initially tried and condemned as a rebel. Thus, military discipline was of great concern for many in the army.

At the end of World War II, the problem of how to deal with the communists also affected the army. Communist resistance fighters were among those who fought to free France, but the same communists had been against the rest of France at the beginning of the war during the time of the Hitler/Stalin pact. In addition, danger from the Soviets in Eastern Europe after 1945 contributed to the psychological strain.

The French war for reestablishing control of Indochina also had a great psychological impact on the army. The turbulent political situation in France contributed to a lack of support from public opinion. In addition, many officers were imprisoned for months in Viet Cong camps and were subject to revolutionary brainwashing.

The psychological warfare problem experienced during the Algerian War, however, was quite different from that experienced in Indochina. In the Algerian conflict, conscripts were used instead of the volunteers used in Indochina. Also, in addition to the close geographical proximity of Algeria to France, there were about 1 million Europeans living there, some families for four or five generations. In both conflicts French officers of the Army and the Navy were sent into villages and small cities to act as administrators and in some cases had to assume the role of police, creating a very difficult situation. Although the military operations were successful on the ground in Algeria, in both the Indochina and Algerian conflicts the international community and public opinion were against them.

STRATEGY AND POLITICS

The mix of civil and military responsibilities and the impact of public opinion demonstrates that psychological warfare can be very close to political issues. The situation in the recent world is one wherein peace and war are not clearly separated nor the enemy always known. In a democracy, military authorities must always be subordinate to civil governments. But a difficulty of the democratic system is that politicians are often able to escape their responsibilities and instead find military leaders to blame.

Psychological warfare is also a branch of general strategy. Strategy is the science for the conduct of action—how to achieve goals, how to attain objectives, and how to win in spite of the opposition of one's adversaries. Qualities of courage, spirit, bravery, domination, and determination are absolutely necessary for any leader involved in a strategic issue. A leader must know his adversary's psychology because the ultimate objective is to conquer the enemy's mind and to dominate him, so the importance of intelligence is paramount. In addition, a good psychology of leadership is necessary, including troop leadership abilities. At least a perception of moral strength is required to appeal to the public and to gain their support.

HISTORICAL USE OF PSYCHOLOGICAL WARFARE

The third point is that psychological warfare is as old as war itself. Historically, one can find many examples of a psychological approach to conflicts. Israeli behavior, for example, is often biblically defined. Among the Chinese, from Sun Tzu to Mao, there is a remarkable continuity of war as the art of deception. From them came the ideas that it is better to defeat the enemy without fighting than to lose people in combat, that you must know your enemy, that spies have many uses, and that the use of mystery and stealth is essential. Today's Chinese military leaders follow the same principles.

There is another tradition which is rooted in the Mongols. It is a set of rules for political and military behavior. Strategic planning was always used in the Mongol armies and intelligence and logistics were of the highest level. Psychological use of terror was also one of their best weapons. In the USSR today, one can find many aspects of the Mongolian culture. The scorched earth strategy and the reliance on brute force to intimidate adversaries come from the Mongol tradition.

Cortez, who with a mere handful of combatants was successful against the entire Aztec empire, presents another example of the many uses of psychological warfare. We find others in colonial traditions, such as those of the British Army in India or in the Middle East. In French history one finds similar examples: admirals of the French Navy in the nineteenth century as well as French generals in the Far East or in Africa were artful in using psychological approaches to emphasize the idea that their purposes were not to kill enemies but to conquer spirits, to honor and respect their adversaries when they had defeated them, and to improve the welfare of and bring civilization to those countries which they had conquered.

Many lessons also can be learned from the study of past and present so-called "low intensity" conflicts. The knowledge and use of psychology to defeat the spirits of the politicians is a proven method, as when Marxists demonstrated how they were able to take control of a group of people with a minority.

Finally, cultural background must be taken into account in any study of psychological warfare. For example, U.S. foreign policy and strategy can often be compared to poker playing, whereas the Soviet Union's strategy is more like that of chess. Americans emphasize the technological aspects of conflicts and try to obtain victory through better technology, whereas the Soviets are still more connected to the principle of mass, surprise, and other more traditional techniques.

PRESENT AND FUTURE USE OF PSYCHOLOGICAL WARFARE

We are living now in the nuclear age and we are living under the effect of the psychology of fear. It is fear and not virtue which has made it possible to avoid a major conflict for the last 40 years. Atomic weapons are not so much combat weapons as political weapons. Remember the political choice made by President Truman during the

Korean War when he decided that it was not possible to use atomic weapons to threaten the Chinese, whereas General MacArthur was still thinking in terms of World War II, believing that when you are involved in a war you are to use all your assets to win. In the nuclear age, the same reasoning used in former wars is not always acceptable.

The East-West challenge has existed 43 years. That challenge is not only between conflicting interests, it is a broad political, philosophical, and ideological challenge, and the champions of the challenge are the two superpowers, the United States and the Union of Soviet Socialist Republics. The main reason for such a challenge is because the values of the two nations are deeply different. There are many pitfalls in this confrontation. The game is not fair, because the Western camp is a coalition of free nations and free nations discuss their problems openly and form coalitions. Historically, it has been easy to fight against coalitions because they have structural vulnerabilities. This is why the Soviet objective has always been, and is still today, to disorganize and to denuclearize the West. Because they know that nuclear weapons are the West's main asset to deter war, this second goal underlies all their tactics, even if their rhetoric changes with the times.

Democracies have many weaknesses when pitted against totalitarian societies, as lies and propaganda are easily launched and used against democratic systems. One not very well known example may be termed the *McCarthy effect*. The negative images associated with the McCarthy era and the House UnAmerican Activities Committee have been used for years by the Left so that it was not possible to say the communists were bad guys because the United States behaved very badly itself in the time of McCarthy. For years it was not possible in the intellectual community to say anything against those Soviet good guys. The communists also get help from the free press. Although the free press is a necessity in a democracy, Westerners easily fall under the influence of words of emotion, often at the expense of reality. The fact is that the press has much power but few responsibilities.

The present danger is that the USSR appears now as a good guy in international society. When there is no more perceived threat—threat being the mother of coalitions—public opinion erodes the necessary efforts for defense. The West is probably facing difficult months and years ahead because the allies have many areas for transAtlantic and European disputes. Those disputes were not decisive so long as there was the threat of Soviet military forces, but it is highly possible that in the months to come economic disputes could launch new difficulties among the allies.

At the same time, the real changes in the USSR require close scrutiny. It is clear that there are real changes within the intellectual circles of Soviet society: today they are discovering how ugly the years of Stalin were and, also, that they are no longer really a superpower. Just two examples from among many others: their health system is very bad, and their environmental problems are huge. One should not be optimistic nor idealistic about changes in the Soviet system; one must be realistic. But it appears that the West is entering a period where it is necessary to keep close contacts with the Soviets because they have the mentality of a besieged fortress. They see the Western military forces as threatening when they look at the map of their country with our task forces encircling them, potentially able to attack the motherland. Given their memories of the "Great Patriotic War" and the pictures showing how horrible it was, they have that kind of psychological basis. Thus, it is necessary both to keep close contact and to be very firm in negotiations with them.

For the future, psychological factors are more and more a matter of global strategy. They interface with the other terms of power and must be integrated into any general strategy. These psychological factors must be fully understood in order to keep stability in the world to show force without having to launch another general war.

9

Contemporary Insurgent Political and Psychological Warfare

Michael Radu

REVOLUTIONARY GOALS AND THE DEFINITION OF THE ENEMY

If it is true that insurgencies succeed or fail politically rather than in the military field, then this is particularly pertinent in the case of contemporary Third World insurgencies. Indeed, even a cursory examination of successful "national liberation" wars—the short term for Marxist-Leninist insurgent organizations in the Third World during the past two decades—suggests that almost all of those that succeeded did so without obtaining a military victory or even stalemate in the field.[1] That was the case in Algeria, Angola, Mozambique, and Guinea Bissau; in Zimbabwe and Aden; probably in Namibia. In fact, in Algeria, Angola, and Namibia the insurgents were militarily defeated at the time of their ultimate victory. In certain instances, such as Cape Verde, São Tomé, and Principe, the "liberators" took over virtually without firing a shot.

In all those instances the insurgents' ultimate victories were the results of political factors: specifically the collapse of the governments' wills to carry on fighting. Where such collapses failed to occur, as in Malaya, Oman, the Philippines during the 1950s, East Timor, and Western Sahara, the insurgency failed decisively. It should be pointed out, however, that in almost all cases mentioned above (the exceptions were Oman and the Philippines) the insurgents' stated goal was national liberation from colonial rule, a goal widely shared by world opinion and the voters in the metropolises. As a result, the decisive political pressure on governments most often came from insiders or from other governments, with the insurgents themselves playing a less important role.

Since the virtual end of anticolonial wars of "national liberation" by the mid-1970s, Third World insurgencies have developed a new character—that of revolutionary wars directed against sovereign national states. In fact, such revolutionary wars were already common in Latin America immediately after the victory of the Cuban Revolution in 1959. The goal now is the imposition by force of a new political, economic, and social order within the national territory rather than obtaining sovereignty.

However, the obvious shift in insurgent goals, from independence to the establishment of a communist system, is more apparent than real. Indeed, most national liberation movements of the 1960s and 1970s, such as the MPLA (Popular Movement for the Liberation of Angola) and FRELIMO (Mozambique Liberation Front), as well as SWAPO (South West Africa People's Organization) and POLISARIO (People's Organization for the Liberation of Saquiet-el-Hamra and Rio de Oro) are today ultimately aimed at establishing a communist system with the end of colonial rule a necessary but intermediate goal. Conversely, all of today's leftist insurgencies or guerrilla umbrella groups, from El Salvador's FMLN (Farabundo Marti National Liberation Front) to the New People's Army (NPA) in the Philippines, and the Peruvian Shining Path,[2] consider imperialism as their main enemy and treat their national governments as, at best, little more than neocolonial puppets.[3] Once imperialist domination is eliminated, they feel socialism will inevitably triumph; hence antiimperialism is the main characteristic of the revolutionary struggle.

The primacy of antiimperialism in the contemporary Marxist-Leninist insurgents' strategy is so obvious as to be shared by such otherwise mortal enemies as Maoists, new-Left and Castroite elements, and old-fashioned pro-Soviet communists. Thus the Shining Path stated that it took up arms "to advance the democratic revolution—to overthrow the exploitation and oppression of imperialism (principally Yankee), bureaucrat capitalism and subsistence semifeudalism."[4] The Salvadoran ERP (People's Revolutionary Army), the FMLN's strongest component, made it clear that "Speaking of our position on the international level means speaking about the struggle against North American imperialism, which we consider the fundamental enemy of all peoples and revolutionary movements."[5] Speaking about their own activities, the Uruguayan Tupamaros declared that "The war will be internationalized, taking a continental character. It will not be determined by ourselves, but by the conditions in Latin America, by its history; it will be determined by the common enemy: imperialism."[6] The ephemeral Guatemalan ultra-leftist group MRP-Ixim (People's Revolutionary Movement-"Food") claimed that "U.S. imperialism . . . caused the economic, political and social crisis that our people are experiencing."[7]

Superficially, the notion that the West in general and the United States in particular could be seen as the strategic enemy even after the end of Western colonial empires—in the case of Latin America more than a century and a half later—seems absurd. It is far less implausible when one examines the roots of such perceptions. On the one hand, as

Marxist-Leninists, the Third World revolutionaries are bound to see the capitalist West as the strategic enemy; on the other hand the conceptual framework of such perceptions is itself provided by Western intellectuals or their pupils. Indeed, nothing has contributed so much to legitimizing hatred for democracy and capitalism as the dependency theory, whose main tenet is that Western economic and, ultimately, political exploitation of the Third World has continued, indeed has intensified, after political independence has been obtained. The proponents of that theory have been either Western scholars and academics or Third World intellectuals trained in the West (Samir Amin, André Gunder-Frank). Moreover, the most persistent, and probably most influential disseminators of it are still Western clergyman, the proponents of liberation theology.[8]

It is impossible to understand, analyze, or gauge the scope and significance of insurgent political warfare today without permanent reference to the entire mental, conceptual, ideological, and emotional universe described above. It is indeed a universal framework of reference, made so by the very nature of revolutionary elites in the Third World—the very group that leads, controls, and organizes guerrilla organizations as well as defines the parameters of their political and psychological operations.

The revolutionary elites are a group that, from Cambodia to Peru, the Philippines to Guatemala and Nicaragua, and from El Salvador to South Africa to Afghanistan, shares similar characteristics. Those include higher education, often in Western institutions,[9] privileged social status, extensive foreign travel, ignorance, and often contempt for national realities beyond elite levels, and a sophisticated understanding of the vulnerabilities, weaknesses, and dynamics of Western democracies.[10]

THE NATURE OF INSURGENT POLITICAL-PSYCHOLOGICAL WARFARE

The previous introduction to the topic of this chapter is necessary because it defines the main target of political-psychological operations—imperialism as represented by the United States—and it suggests the methods to be pursued. The latter, following the Vietnamese, Nicaraguan, and to some extent Cuban examples, suggests a concentration of effort "in the belly of the beast," or inside Western democracies. Indeed, if the main enemy is imperialism, political action at the center is decisive. Popular mobilization at home, important as it is, is only effective insofar as it is nationalist, that is, anti-Western, or, more often than not, anti-American.

It is important, at this point, to clarify a number of terms and concepts. The terms political and psychological warfare tend to have somewhat different meanings in the case of irregular military forces, both in terms of their importance and content. In traditional and conventional terms, political and psychological warfare are intended to supplement and enhance the effectiveness of military operations after the outbreak of hostilities. Their target tends to be the weakening of the enemy's will to persist in fighting and that of the morale of enemy troops in particular. Ultimately, however, the resolution of the conflict is expected to be obtained in the field. In an insurgency context, however, military operations are more often than not seen as a supplement to political warfare. It is significant in this respect that the most serious accusation against a rival group on the left seems to be that of militarism, of minimizing the political aspects of revolutionary struggle to the advantage of pure violence. A good example is provided by the 1978 criticism of the Salvadoran ERP by El Salvador's oldest guerrilla organization, the Popular Liberation Forces (FPL): "The ERP's focus on the revolution was purely military. It was more concerned about technical combat aspects than about political military structure. . . . Harsh realities led the ERP to a new turning point, this time in the direction of crude adventurist militarism."[11] Similar accusations from within the revolutionary Left were made against the Guatemalan ORPA (Revolutionary Organization of People in Arms). On the other hand, strict reliance on political action, at the cost of quasi-total neglect of military operations could lead to accusations of "parliamentary cretinism" and "rotten electoralism" by groups like Peru's Sendero Luminoso.[12]

Despite such extreme positions, it is highly significant that all revolutionary organizations of the Left today define themselves on paper as political first and military second, with the term "politico/military" being the most often used. In fact, even Sendero claims to be a party, that is, a political organization. The implication is that regardless of the importance given to armed struggle, the main object is *political* power, under *political* leadership, with *political* aims. It is indeed the primacy of politics, the understanding of their decisive role that distinguishes genuine insurgents from mere terrorists. It is also the importance of politics that makes political warfare a vital part of insurgent strategy in most cases.

POLITICAL WARFARE AND PROPAGANDA

The supremacy of politics in revolutionary warfare is a characteristic shared by all Marxist-Leninist groups today and, in fact, since Lenin himself urged professional revolutionaries to act as "Theoreticians, propagandists, agitators, and organizers." The terms propaganda and agitation became coterminous after 1917, and were since used jointly as "agitprop."[13] Furthermore, many insurgent organizations fighting today against Marxist-Leninist governments are led by assiduous students, or even former supporters, of Leninism or Maoism; thus a similar approach to political warfare and propaganda can be detected in their cases. That is particularly clear in the cases of UNITA and its leader, Jonas Malheiro Savimbi; of Ahmad Shah Massoud in Afghanistan's Panjshir Valley; of Afonso Dhlakhama of Renamo in Mozambique; and obviously in those of the Khmer Rouge in Cambodia, the EPLF in Eritrea, and the Tigre People's Liberation Front in Ethiopia.[14]

Generally, in most contemporary insurgencies whether Marxist-Leninist or anticommunist, politics, propaganda, and political warfare are not only equally essential, but they also have similar or at least closely related characteristics, follow similar methods and principles, and aim at similar results.

Conceptually, the most important characteristics of today's insurgencies' approach to political warfare could simply be described as follows:

- Since revolutionary warfare is about changing the nature of the entire society, in all its aspects, and all warfare is ultimately political, all aspects of the society are part of politics and therefore subject to political warfare (hence, for instance, the treatment of refugees);

- Political warfare is ultimately *propaganda*, hereby defined as a coherent, consistent and conscious attempt to manipulate facts, images, perceptions, and emotions. Since politics are political warfare, and political warfare is propaganda, the entire political behavior of insurgents can only be understood as propaganda or, in the Leninist tradition, as agitprop.

Once the above elements are accepted, the entire behavior of communist and, to some large but more limited extent,[15] anticommunist insurgents in Latin America, Africa, and Asia could and should be seen as part of their agitprop techniques, structure, slogans, language, targets, and goals.

TECHNIQUES AND METHODS

To some extent, particularly in domestic matters, there is a significant difference between propaganda and psychological warfare, and the methods of pursuing each are not only different but could even cancel each other out. Furthermore, the gap between the language and proclaimed goals of international propaganda and those of domestic agitprop also vary greatly. In both instances the essential variable is the nature of what is generally called propaganda by deed or armed propaganda.

Propaganda by Deed

A general definition of the term would define *propaganda by deed* as any violent action (hence the synonym, armed propaganda) aimed at demonstrating the military strength and seriousness of purpose of the insurgents. Carlos Marighella, widely recognized by such different groups as the Tupamaros, Montoneros, FMLN, the Italian Red Brigades, the IRA, and the German Red Army faction as the main theoretician of contemporary urban guerrilla warfare, defined *armed propaganda* as "an integral part of coordinated urban guerrilla operations. Each individual urban guerrilla operation is in itself a form of armed propaganda."[16] The term generally includes a strong accent on intimidation and, since it requires publicity to be effective beyond its immediate location, it is highly dependent upon media coverage.

The main characteristic of any propaganda by deed action is its violence; hence, most such actions involve kidnapping, murder, or any related activities, generally but often misleadingly defined as terrorism. Unlike true terrorism, which aims at sheer intimidation and targets victims with no involvement whatsoever in the conflict, whether old Americans in wheelchairs, travelers on an intercontinental flight, or just passersby,[17] terror as a method of propaganda by deed is specifically and "reasonably" targeted to persons who are, albeit marginally, involved. The following brief analysis of the main means of propaganda by deed clearly demonstrates the distinction.

Kidnapping

Kidnapping is almost universally seen by insurgent organizations as a most effective propaganda instrument. Marighella stated that "Kidnapping is used as a means of exchanging or liberating imprisoned revolutionary comrades, or forcing the military dictatorship to end the torture of prisoners. Kidnapping of non-political national figures, such as well known artists, sports personalities, or publicized individuals in other fields, may be used as a form of revolutionary propaganda. It is important that these forms of kidnapping encourage public sympathy and acceptance."[18] The Tupamaros, probably Marighella's closest disciples, described kidnapping as a limited form of urban guerrilla warfare, a "modest proof of the incalculable possibilities offered by the revolutionary prison, where one could indefinitely detain regime personalities, hatchetmen of the repression, foreign representatives and generally key personalities of the regime."[19]

The method and practice of kidnapping took a new dimension in El Salvador during the late 1970s and involved political and financial aspects as well as larger international ones. Thus, the FPL, until the early 1980s the largest and most influential FMLN element, claimed that kidnapping "is a political, economic, and military weapon which the proletariat and the people in general use for the defense of their interests. It is an instrument that serves the revolutionaries in order to win well defined objectives otherwise impossible to reach. From this point of view, kidnapping has economic, political, and military objectives."[20] At the same time ERP, now the dominant element of the FMLN, saw kidnappings as "acts of revolutionary justice. . . . Therefore, an operation during which the revolutionary forces capture, aiming at justice, one of the members of the families that possess the economic and political power on the basis of exploiting the popular sectors, becomes a political instrument that unveils the structure of injustice and class polarization that dominates the country."[21]

In many instances the kidnapping of prominent personalities or even of their relatives is not only a propaganda by deed action, but serves to reinforce other aspects of a guerrilla group's political warfare operations. The best and most frequent example is the kidnappers' requirement that their propaganda be published and broadcast at government expense. That was the case with the September 1985 kidnapping of President Napoleon Duarte's daughter, Inez, by a Communist party commando. In the case of the 1982 kidnapping of the daughters of Presidents Suazo Cordoba of Honduras and Rios Montt of Guatemala, the group involved, the MRP-Ixim, not only had its propaganda published in exchange for their release, but it actually made its debut on Guatemala's revolutionary scene. Significantly, the MRP-Ixim demanded that its manifesto be published both in Guatemala and abroad.[22]

The preoccupation with external propaganda and its links to the practice of kidnapping were best described by a Salvadoran FPL leader, when she stated that "Kidnapping, as a political objective, also allows us, at a given moment, to manifest our solidarity with the struggle of other peoples and, at the same time, denounce some imperialist maneuver. This is the case with the South African ambassador Archibald Gardner Dunn. Salvadorans feel a profound love and enthusiasm for the struggle of the peoples of South Africa, Namibia, Rhodesia, and Palestine against racism and its policy of apartheid."[23]

Another aspect of the linkage between kidnapping and external propaganda is the use of foreign hostages to gain formal or actual political recognition from the hostages' governments by conditioning release upon involvement of diplomats in the process. UNITA in Angola has perfected this type of kidnapping for external propaganda purposes, as demonstrated by a number of incidents. In March 1983, UNITA forces captured 66 Czechoslovak technicians and their families at the town of Alto Catumbela in Western Angola. After a forced march of almost two months across the entire country, during which the Czech government put pressure on the Cubans and the Luanda regime not to attack the UNITA column—a highly embarrassing request for the prestige and image of both Cuba and Angola—the hostages were released. The release, which involved help from South Africa—yet another embarrassment for Prague—was witnessed by international media, involved appeals from the pope, and generally provided a huge propaganda bonanza for the guerrillas.[24] In December 1984, a UNITA attack against the diamond mines of Cafunfo resulted in the capture of a number of British, American, and Filipino citizens. Once again, after a long march across the country during which the Angolan government was again proven unable to stop the insurgents, the hostages were released from UNITA's capital of Jamba amidst an international media gala. The BBC and ITN from Britain, and Dutch, French, West German, and Italian TV crews were at hand, all in effect providing massive, cost free, and generally positive coverage of UNITA.[25]

Not all kidnappings have such complex or high-minded goals, not even in the rhetoric of the kidnappers. For instance, the Guatemalan FAR (Rebel Armed Forces) group admitted that, in cooperation with the Salvadoran FPL, "We accept responsibility for the execution of the millionaire José Ernesto Regalado Duena. The latter chose not to remit the tax that the revolutionaries of the brother country established to finance the revolutionary struggle, and incited his family to collaborate with the repressive forces of El Salvador."[26] The reason for the kidnapping and subsequent murder was clearly financial rather than political, FAR's feeble attempts to cover those actions with revolutionary rhetoric notwithstanding. Nevertheless, some kidnappings have brought so much money to certain insurgent groups, particularly the Argentine Montoneros and the Salvadoran FARN (Armed Forces of National Resistance, a FMLN group) that they were able to engage in highly expensive propaganda campaigns,[27] involving paid advertisements in internationally known journals and on radio stations.

Political Murder

Like kidnapping, political murder is part and parcel of guerrilla politico-psychological warfare; unlike it, however, it reaches the wealthy and the poor alike, has a stronger psychological component, and is generally limited to the domestic audience rather than the international arena. Nevertheless, a major type of assassination clearly intended as a propaganda action and targeting a foreign audience is the murder of foreign diplomats, a rather common occurrence in Latin America in the late 1960s and early 1970s, from Guatemala to Uruguay and from El Salvador to Argentina. The goals of such actions are the intimidation of a foreign government seen as supporting the target regime and, equally if not more important, massive publicity in foreign media.

The same reasons could be found behind the assassination of foreign citizens in general; the murder of four U.S. Marines in San Salvador in June 1985 by the PRTC (Revolutionary Party of Central American Workers, one of the FMLN groups) was openly intended to help create yet another Vietnam psychosis in the United States, in support of the FMLN front's propaganda within the United States. Conversely, in order to encourage those in the United States inclined to take the latest FMLN peace proposals seriously, in February 1989 the FMLN suspended their attacks against U.S. facilities.[28]

One of the least noticed but extremely important motives of political assassination is its role of keeping past insurgent activities alive in public memory and projecting an image of the revolutionaries as having a long memory and inexorable purpose.

Specifically, this refers to the murder of people long retired from politics or the military, but accused by the insurgents of past crimes of the people. In many instances this type of action, so far restricted to Latin America, serves either to launch a new revolutionary group into notoriety or to keep small, marginal or largely inactive groups in the public eye.

The first scenario certainly applied to the June 1970 assassination in Buenos Aires of retired General Pedro Eugenio Abramburu by the then almost unknown Montoneros. The victim's crime was to have led the 1955 coup against Juan Domingo Peron, and it instantly made the Montoneros famous, at least in Argentina. Most recently, on January 25, 1988, the Honduran Cinchoneros (Popular Liberation Movement--Chinchoneros), a 1981 creation of the Salvadoran ERP[29] which was widely seen as defunct by 1986, killed retired General Gustavo Alvarez Martinez, former defense minister, and promptly captured the headlines in Honduras and the United States.[30]

The second scenario, the killing of old men in order to prove a group's continued dedication and genuine revolutionary credentials when those are seriously doubted on the Left, occurred in Guatemala and El Salvador. In Guatemala, the PGT (Guatemalan Party of Labor, the pro-Soviet communists) has seldom been involved in violence in the past 15 years. Nevertheless, it claimed responsibility for the killings of Colonels Rafael Arriaga Bosque in September 1977 and Maximo Zepeda Martinez in March 1980. Both were long retired and both were involved in the late 1960s counterinsurgency campaign that destroyed the Guatemalan insurgencies at that time, Arriaga as defense minister and Zepeda as police commander.[31] In El Salvador a splinter of the FPL, the Clara Elisabeth Ramírez Metropolitan Front (CERMF) still loyal to FPL's founder Salvador Cayetano Carpio, was heavily involved in 1985 in a campaign of murders of old officials. The campaign carried the general and highly instructive label of "Long Live the Fifteen Years of Prolonged People's Warfare." Among the victims were a former policeman, Manuel de Jeses Flores Mateo, who allegedly tortured Carpio during the 1950s, and a retired general, José Alberto "Chele" Medrano, a hero of the 1969 war with Honduras and presidential candidate in 1974. Most importantly, Medrano was the founder of the paramilitary, mass-based ORDEN (Nationalist Democratic Organization), which effectively prevented the formation and development of leftist insurgent groups in El Salvador during the 1960s.[32]

For the more established insurgent groups, which do not have to prove their strength and commitment or trumpet their entrance on the scene, the spectacular political assassination tends to involve a combination of propaganda and practical considerations. The failed September 1986 attempt by the BPMR (Manuel Rodriguez Patriotic Front, the armed branch of the Chilean Communist party) to assassinate President Augusto Pinochet is a good example. From its code name "Operation Twentieth Century" and the involvement of over one hundred cadres, it was clearly intended as a major propaganda coup; and, in light of the Pinochet regime's highly personalistic nature, it aimed at producing a change of government.[33] The CERMF also murdered Colonel Aristides Cienfuegos, the army's press spokesman, presumably because of his effectiveness in presenting the government's side of the war. This was an interesting instance of a propaganda killing directed at a chief propagandist of the enemy. Similarly, the murder of three national police chiefs in two years by the Argentine Montoneros or that of army spokesman Aristides Cienfuegos by the CERMF in El Salvador in 1985 had both practical and propaganda merits from an insurgent viewpoint. The killing of Colonel Cienfuegos, clearly the most articulate and engaging public relations expert of the Salvadoran military in many years could be seen as an almost pure example of a propaganda war solved by violent means.

High level assassination could be intended as preventive when the victims are personalities perceived as potentially strong would-be enemies of the revolution. The best examples are provided by the Sandinista regime in Nicaragua, relevant to this chapter in light of the previously insurgent nature of the now-ruling FSLN (Sandinista Front of National Liberation). Once in power, the Sandinistas engaged in a housecleaning operation, aiming at removing any possible focus of armed or political opposition. The campaign was led by the then and now National Directorate member and interior minister, Tom S. Borge and was often executed by his sidekick Lenin Cerna, today deputy interior minister. The concrete results were the physical removal (murder) of potential military, foreign, and domestic leaders and symbols, including Major Pablo Emilio Salazar, "Comandante Bravo," the only proven, popular and competent officer of Somoza's National Guard; Jorge Salazar, a businessman, the most charismatic, politically shrewd and charismatic leader of the internal opposition; and Anastasio Somoza himself. All three were killed by or through the agents of Nicaragua's secret intelligence. "Bravo" was murdered in Tegucigalpa after long torture by Lenin Cerna, Jorge Salazar was set up by the Sandinistas secret police, and Somoza was blown up in Asuncion by the remnants of the Argentine ERP.[34]

Another type of spectacular assassination, that of defectors, has as its primary goal the intimidation of those defectors still active as well as the prevention of such actions in the future. Two spectacular examples of such actions are the murder of former Communist party and African National Congress (ANC) leader Bartholomew Hlapane in Johannesburg in 1984, following his devastating testimony to the U.S. Senate, and, more recently, the FPL killing in El Salvador of Napoleon Romero (a.k.a. Miguel Castellanos) on February 17, 1989. In the latter case, according to the FPL's Radio Farabundo Marti, "the execution not only represents a military victory of the people but a political blow against the traitors and the supporters of the dictatorship. . . . Now, the traitor is where he belongs—in the garbage dump of history."[35]

The goals of assassination, or revolutionary justice in guerrilla language, are numerous and complex, but intimidations and the creation of an illusion of insurgent omnipresence and omnipotence are the dominant elements, together with population control.

Routine Terror

In number of victims and in long-term impact, however, assassinations are not the most important aspect of the insurgents' use of murder as a propaganda weapon. That role is clearly played by the killing of countless civilians as part of a massive and quite effective campaign of intimidation, population control, and recruitment.

A number of distinctions have to be made in this respect. First, it would be misleading and probably counterproductive to confuse the systematic use of terror tactics by the insurgents, particularly in the countryside, with the nature of such organizations as the FMLN, the URNG, the NPA (New People's Army in the Philippines), or the Khmer Rouge, and even more so in the cases of UNITA in Angola, the Contras in Nicaragua or Renamo in Mozambique. *Terror* is a necessary part of any guerrilla campaign; *terrorism* is a goal, per se, and at best a war in and for the media; in other words, terrorism is little more than a deadly theater, whereas insurgency and terror are natural parts of a revolutionary process, whether directed against democratic, pro-Western governments or against communist regimes. Second, anticommunist insurgents, particularly those supported by the United States within the loose framework of the Reagan Doctrine, are generally reluctant to engage in large-scale, routine terror against civilians, although UNITA, the Mujahedeen, and the Contras have all used various degrees of intimidation against noncombatants and have occasionally killed infiltrators and enemy cadres.[36] Furthermore, no spectacular assassination of a government official, civilian or military, national or foreign, was ever plausibly attributed to anticommunist guerrillas.

It is only against this background that one could understand statements like this from Joaquan Villalobos:

> We do not intend to allow the Army to wage war and terror against the people with impunity. The war will also reach residential areas to influence groups with decision-

> making power. Perhaps this will make them consider other solutions and alternatives. Not long ago these groups were claiming the war is over. It is not; on the contrary, the first phase of our urban activities is just beginning.[37]

The aim of such a statement, promptly reinforced by the FMLN's burning down a coffee-processing plant owned by Alfredo Cristiani, the leader of the conservative ARENA party and president of El Salvador, is to instill terror in the hearts of guerrilla opponents, to intimidate and cow them. Threats, murders, and intimidation against civilian elected officials at the local level are, in all contemporary Marxist-Leninist insurgent campaigns, as common as they are effective. Once again, the Salvadoran FMLN is the model, although far from being the first, the only, or even the main practitioner of such tactics.

As far as the FMLN is concerned, mayors elected during the 1988 municipal elections are nothing but local representatives of the Proyecto, the grand name the guerrillas give to the Salvadoran Army's counterinsurgency strategy attributed to the United States. Fighting the Proyecto is fighting against imperialism as well as against the oligarchy. Therefore, all means, particularly intimidation and murder, are legitimate, since the real prize is population control. So, whether they are left-of-center Christian Democrats (whose statist, anti-free enterprise policies during the past five years have contributed heavily to El Salvador's economic problems), or bona fide conservatives like Alfredo Cristiani, the mayors are the FMLN's enemy and have been subjected to a campaign of murder and intimidation.

Nor is the FMLN bashful in threatening mayors. A Radio Venceremos broadcast on February 19, 1989, stated that "We remind them (the mayors) that they must publicly announce their resignation within 72 hours. This specifically applies to the mayors of Joateca, Perquin, Torola and San Fernando. . . . No trace of the dictatorship's municipal governments, which the army is not capable of defending, can remain."[38] The aims of the FMLN were as clear as they were successful: (a) eliminate government institutions from contested areas; (b) discredit the importance of elections and elected officials; and (c) demonstrate FMLN power in certain areas, and make the government look ineffective and incapable of protecting the population. The killing of mayors, 12 of whom were murdered by the end of January 1989, resulted in 106 out of 262 elected officials resigning from their offices.[39] The result was an absence of elected party officials able to campaign for their respective parties, and a collapse of civilian government presence in most of eastern El Salvador.[40] Many of the mayors threatened by the FMLN were women, often lower-class supporters of ARENA, such as Marta Gomez, mayor of Cojutepeque and daughter of the cook of Ruben Zamora's family; Zamora was the oligarch representing the FDR, FMLN's political arm.[41]

TERROR AND THE EVOLUTION OF INSURGENCIES

The use of political assassination and terror as instruments of population control are not, generally speaking, equally important at all stages of an insurgency, from start to victory or decisive defeat. Sometimes both methods are used at the same time, as they are today in Peru by Sendero Luminoso and in El Salvador by the FMLN; in each case the insurgents are mature, well established organizations operating in the cities as well as in the countryside. More often, however, the emphasis falls on one or the other of the two instruments, depending on the stage of development of the insurgent organization.

As a general rule, the use of spectacular assassinations is more frequent at the start and end of a failed insurgency, when the need for publicity is the most important consideration. Well established and large insurgent groups tend to avoid most types of spectacular murders, with the exception of defectors, and particularly those of foreigners and diplomats. The reason is that at that stage the revolutionaries' main propaganda goal is to influence international opinion and to consolidate and establish themselves as legitimate parties in a civil war. In such circumstances the assassination of major figures, at least when publicly admitted, is counterproductive.

On the other hand, the use of terror against the population at large, particularly in the countryside, tends to be routinely employed by these large and established insurgent groups, rather than by newly established or declining ones. A new or declining insurgent movement needs voluntary support from the population, especially in order to grow or revive its fortunes, and thus it can ill afford indiscriminate terror against civilians. Large organizations, on the other hand, need to control large numbers of civilians in their operational and rear areas, and to weaken the communities' will to resist in other areas. For these purposes terror against civilians is the most effective instrument.

AGITPROP—THE "NON-VIOLENT" ASPECTS

While there is no detectable distinction between pure violence and pure propaganda in the strategy of contemporary insurgent organizations, there are some distinctions to be made in regard to organization, methods, and goals of propaganda experts and military commanders. These distinctions may not always be obvious, either because the leadership is centered in one person, who controls both political and military organs of the movement (UNITA, Sendero), or because propagandists are also military commanders (ERP in El Salvador). Nevertheless, the difference in approach is often clear and always institutionalized within a given guerrilla group.

The most important, indeed essential, goal of any insurgent group's propaganda is to capture the language regarding itself and its foes and hence the image-making mechanism in both the media and the public opinion.

REVOLUTIONARY LANGUAGE AND PROPAGANDA

The Left's pirating of public language is neither limited to its peculiar definition of essential concepts nor is it restrained by local understandings of revolutionary rhetoric. On the contrary, it is directed to transforming the very substance and outcome of public debate, since terms such as peace, democracy, and justice are consistently used as if they were similar to their meaning in the West. Such brutalization of language is often playing upon U.S. libertarian mentality and open society by using terms like human rights, democracy, people, and independence to describe guerrilla goals; using symbols like Jefferson, or terms like the American Revolution as false analogies to insurgent actions; and using media-reinforced negative images and slogans like Contras, Somocistas, South African puppets, and the return of Pol Pot. By contrast, guerrillas describe themselves as peace loving, patriots, always ready to negotiate or, as in Guatemala and El Salvador, prepared to humanize the war. In both cases the dehumanizing aspects of the conflict, that is, its threat to civilians, is largely the result of the guerrillas' extensive and purposeful use of mines intended to maim rather than kill.

Among the best examples of effective guerrilla misuse of an important concept is the almost general description of insurgent assassinations by the Western media as executions. This blurring of the essential distinction between an act of justice following due process of law and an instrument of terror serves to give the insurgents a degree of legitimacy they have done little to deserve.

Probably the most spectacular victory of a Marxist-Leninist organization, albeit one already in power rather than seeking it, was the Sandinistas' ability to define the armed opposition to their own rule as counterrevolutionaries, or simply Contras. In light of the Western, and particularly U.S. public's perception of revolution as inherently positive and progressive, being against revolution, that is, the FSLN regime, implicitly means being retrograde, reactionary, or, as the Sandinistas have consistently and absurdly claimed, being a follower of the Somoza regime.[42] The fact that even former president Ronald Reagan claimed to be a Contra was not a victory for the democratic resistance in Nicaragua but, ultimately, for the Sandinistas, inasmuch as they and their Western supporters have coined the term.

Human Rights as Political Warfare

The Sandinistas' success in mislabeling, and thus delegitimizing their foes is an example of government counterinsurgent propaganda, but such propaganda techniques have long become a staple of insurgents as well. Among other examples of such hijacking of once meaningful and legitimate terms are the use of words such as repression and violations of human rights. In El Salvador the FMLN has from the time of the Carter administration described any government attempt to control or suppress violent activities by guerrilla fronts, particularly in the capital, as repression. Although FMLN commander Marisol Galindo stated that "the regime and the U.S. government devised a strategy based on genocide," she also admitted that claims of human rights violations (by the Carter administration) were instrumental in bringing about the 1979 coup in El Salvador, which "proves that reports and resolutions by international organizations do exert pressure on governments."[43]

How and why some human rights organizations support such distortions of normal language and confuse counterinsurgency and the government's right and duty to maintain law and order with violations of human rights has been extensively documented.[44] What has been less well documented is the enormous extent to which insurgent groups, particularly those on the Left, have used the issue of human rights to their advantage, especially since that topic was made so prominent during the Carter years. Indeed, virtually all Third World human rights organizations are left of center in their orientation, as are a majority of their Western counterparts. In Latin America in general and in Central America in particular (Nicaragua being the only partial exception[45]) the mushrooming human rights groups are little more than fronts for the guerrillas. Such groups include COMRADES (Committee of Mothers and Relatives of Political Prisoners and the Disappeared), a group founded and controlled by FARN cadres from the start and which, in addition to being used for propaganda purposes, collects funds for the FMLN in the United States; the "Marianella Garcia Villa" Committee for Freedom of Political Prisoners and Missing People, named after the first leader of the self-appointed Committee for Human Rights in El Salvador, killed while traveling at night with a guerrilla column, and many others.[46]

The sheer number of such pseudo-human-rights organizations,[47] with their uniformly one-sided reports, gives unaware outsiders the false impression of a majority opinion against the government. In addition to their propaganda value internationally, often enhanced by their real or alleged association with religious institutions, such groups also serve for fund-raising purposes abroad, and support military objectives at home. Regarding the latter aspect, human rights groups in South Africa and El Salvador are outspoken against the military draft—aiming at disarming the government—while in Namibia, Peru, El Salvador, and the Philippines, they noisily protest the use of aircraft against the guerrillas, which they routinely describe as indiscriminate bombing of civilians.

Finally, guerrillas throughout the world, have perfected the art of using refugees for propaganda, fund-raising, and logistical purposes. The Western sympathy naturally extended to refugees is manipulated into obtaining medicines and food for combatants, and refugee camps are often transformed into rest and recreation areas for insurgents. The FMLN has transformed refugee camps in Honduras into a laboratory for indoctrination, de facto liberated areas, and sources of supplies, all under the largely indifferent or sympathetic eyes of international relief organizations.[48]

Guerrillas are perfectly aware of the difference between their own human rights practices and those of the governments they fight against. As a spokesman for the Democratic Convergence, the legal front of the FMLN's civilian arm, the Revolutionary Democratic Front described it:

> The human rights situation is a different problem because the Salvadoran Government has signed international agreements on human rights, not the guerrillas or the rebels . . . we have stated that the Salvadoran Government is exclusively responsible for human rights in this country.[49]

This asymmetry between the obligations of governments and insurgents is a major difference between the ability of anticommunist and communist contemporary insurgents to use the topic of human rights as an effective propaganda instrument. Marxist-Leninist insurgents, fighting pro-Western regimes, are always able to take advantage of those regimes' inhibitions, in many cases the result of open pressure from Western governments. Congressional human rights strings have always been attached to U.S. aid to El Salvador or Guatemala, to name just two cases; no such conditions were imposed by the foreign supporters of those governments. Conversely, Western aid to anticommunist insurgents in Angola or Nicaragua was also tied to their respect for human rights;[50] no such restraints are or could conceivably be imposed on the governments they fight against. The result is that one of the contemporary insurgents' most valuable propaganda tools is denied to pro-Western guerrillas. The same applies to another, perhaps even more important instrument of political warfare, the use of legal fronts, both at home and abroad.

Political Fronts and Political Warfare

The use of fronts—ostensibly peaceful, legal, or semi-legal institutions—for the purposes of illegal, revolutionary movements aiming at the destruction of the very institutional structure of a country is at least as old as Leninism and probably older. As often described, fronts or, in Comintern parlance, united fronts, could be established on the basis of a tactical alliance with noncommunist parties or organizations (united front from above) or through communist infiltration and subversion of legitimate groups (united front from below or, in Trotskyite terminology *entryism*). A third and fourth type of front have been often used by insurgents in significant addition to older techniques. These are organizations entirely created by guerrillas and international fronts. Among the former are union federations and umbrellas and human rights groups. Previously the latter were established and controlled by communist governments only; the FMLN in El Salvador was the first to perfect them under nongovernmental control.

It is impossible to overestimate the importance of guerrilla fronts for insurgent political-psychological warfare. They divert scarce government resources, economic and political, from building an antiinsurgent national consensus as protection against economic and social warfare.

Insurgent-controlled but legal media engage in open propaganda against the military and civilian authorities.[51] Guerrilla offensives are supported by illegal and often violent strikes by the unions they control—a pattern common in El Salvador since the mid-1970s, while sympathetic and prominent clergymen call for a stop of external aid to the embattled government and for its soldiers to lay down their weapons.[52] University and high school students are organized by professional revolutionary cadres, and campuses are transformed into recruiting grounds, training camps, and arsenals for the guerrillas.[53] Government attempts to control or suppress such activities offer the Left an infinity of propaganda bonanzas and badly damage the regime's international image, particularly in the West. It is this latter aspect, the fronts' ability to influence public opinion abroad, that is perhaps the most important aspect of political and psychological warfare in an insurgency context.

External Fronts and "Solidarity" Groups

In 1975 Joaquan Villalobos, the FMLN's top leader, murdered with his own hands one of his revolutionary rivals, Roque Dalton, an act for which he was publicly condemned at the time by the leaders of both FPL and FARN; in 1980 he was probably behind the mysterious murder of FARN founder Ernesto Jovel.[54] In October 1988 the same Villalobos was successively received by the democratically elected presidents of Costa Rica, Ecuador, Peru, and Uruguay; and at the beginning of 1989 he published his views on the Salvadoran situation in the prestigious *Foreign Policy* magazine. What has transformed the most violent leader of a totalitarian organization, one personally accused of unjustified murder of his own comrades, into an acceptable interlocutor of nontotalitarian chiefs of state?

The answer may well be that provided by Villalobos himself in 1983: "We have a large solidarity apparatus that encompasses the entire planet, even in the United States, where one of the most active centers of solidarity exists."[55] Indeed, in one of the largest and most effective propaganda campaigns ever seen since the Vietnam War, the FMLN has succeeded in transforming its and Villalobos' image from that of bloodthirsty totalitarians[56] into acceptable partners for a share of power in El Salvador. Pro-FMLN demonstrations, teach-ins, friendly academic and media coverage, church-sponsored propaganda throughout the West have become an avalanche of disinformation which is now bearing fruit. Nor was all this the result of any spontaneous sentiment in the United States, Australia, or Western Europe that the FMLN cause was self-evidently just. On the contrary, it was manipulated all the time from a central propaganda headquarters, largely dominated by ERP, itself. The largest U.S. front for the FMLN, CISPES (Committee in Solidarity with the People of El Salvador), is part of the International Committee of Solidarity with the People of El Salvador (ICSPES) and in control of most pro-FMLN activities. Both were established in the United States by Farid Handal, now FMLN representative to Havana and brother of Shafick Jorge Handal, the general secretary of the CPES.[57]

Although no contemporary insurgent organization has reached the FMLN's skills and success in establishing such enormous and effective international propaganda fronts, many have tried. The Guatemalan ORPA claimed that it "has an international front which, in addition to publicizing [ORPA's] work maintains close ties with international revolutionary organizations."[58] The Communist party of the Philippines and its armed branch, the New People's Army (CPP-NPA), have established an active network of solidarity groups in Western Europe, led by an ex-Catholic priest, and engaged in both fund-raising and propaganda.[59]

Once again, as in the case of the human rights issue, Marxist-Leninist insurgent organizations have a decisive advantage over anticommunist guerrillas. Not only can anticommunist guerrillas not establish legal fronts or media outlets at home, for the obvious reason that totalitarian regimes do not allow such independent institutions to exist, they are also handicapped abroad.

Indeed, while there are a few national and international organizations that are dedicated to supporting anticommunist causes, none has the resources available to the Left and, particularly important, none has official or even significant and consistent covert financial support from governments: there are no conservative Cubas, Libyas, or Soviet Unions. The result is that anticommunist insurgents have to limit their activities abroad to public relations[60] and lobbying rather than the mobilization of masses of professional activists. Some, such as UNITA, are very successful; others (the Mujahedeen) enjoy widespread support because of the obvious justice of their cause rather than as a result of their propaganda skills; still others are totally inept and lack resources (RENAMO) or have been poorly advised by their supporters (the Contras). In addition to the difficulties inherent in the conduct of internal propaganda against Marxist-Leninist regimes, many anticommunist insurgent groups also lack the qualified personnel, techniques, and resources needed for massive and protracted political warfare campaigns.

Organizational Structure and Technical Means

No matter how sophisticated an insurgent organization's understanding of the necessity and importance of political warfare, or how just and self-evident the rightfulness of its cause, it still requires a number of fundamental organizational and technical means and skills. The availability of funds is also an important factor in the successful conduct of political warfare, although far less so than the willingness to allocate even scarce resources to such purposes. It is significant that some anticommunist organizations have provided virtually no funds to their missions and operations abroad (the Mujahedeen, RENAMO, the Cambodians), often forcing them to rely on external contributions. Some such contributions may come with strings attached, while some are totally insufficient for successful operations. In other cases, prestigious, expensive public relations firms were retained, although their understanding of the problem was, at best, related to lobbying and placing friendly pieces in the pages of newspapers.

None of these methods could compare with the intensity, efficiency, and effectiveness of native professional propagandists and leftist fanatics available to Marxist-Leninist insurgent groups. With the exception of UNITA, no anticommunist guerrilla organization has large numbers of influential, well-educated, disciplined, and dedicated full-time professional propagandists abroad. By contrast, the FDR alone includes members of the Socialist International, all trained in the West and very well connected, people like Guillermo Ungo, Ruben Zamora, Hector Oqueli and many others. In addition, the Mexico City-based FDR also includes direct and very powerful representatives of the FMLN guerrilla groups themselves. Similarly, at the time the Sandinistas were fighting against Somoza, their most effective front, the "Group of Twelve," included U.S.-trained Catholic priests and respected businessmen.

The importance of political warfare in the internal guerrilla structure is also far greater among Marxist-Leninist groups than among their anticommunist counterparts. For instance, three of the seven members of the Salvadoran ERP's Political Commission, all women, are engaged full-time in political-psychological warfare.[61] The founder and former leader of the CPP-NPA, Jose Maria Sison, while claiming to be outside the party now, is active in conducting proguerrilla propaganda. By contrast, only UNITA among anticommunist insurgencies has allowed its external spokesmen any significant say in strategy.

In light of the high rates of illiteracy characteristic to most Third

World countries, all insurgent groups (with the partial exception of the Mujahedeen[62]) stress radio and visual distribution of propaganda rather than publications. Almost all have their own radio stations, in many cases (El Salvador, Mozambique, Cambodia, and Nicaragua) covering the entire national territory. Visual propaganda, including leaflets and graffiti, are widely used, mostly as a result of the activities of sympathizers and front organizations members. In Guatemala and El Salvador the insurgent umbrellas have even established their own press agencies (Cerigua and Venceremos, respectively), supplying a steady stream of disinformation. In addition, Marxist-Leninist groups, but not their anticommunist counterparts, have regular and free access to the official media of friendly regimes, such as Radio Havana and Radio Sandino, *The World Marxist Review*, *Granma* and *Barricada*.

CONCLUSION

For reasons of space, and considering the complexity of the topic, the picture of insurgent political-psychological warfare depicted in this chapter is far from complete; nor are all the techniques used or available mentioned, let alone analyzed. The means available to governments to counter guerrilla propaganda are barely mentioned, as are the many examples of ineffective propaganda campaigns. Nevertheless, one fundamental lesson seems to be evident: Not only is contemporary insurgent political-psychological warfare essential for the outcome of the growing number of Third World civil wars, it has reached a level of sophistication that requires substantially more attention, understanding, and resources than most Western or pro-Western governments have so far been able or prepared to invest.

NOTES

1. Very few spokesmen for "national liberation" movements would admit this fact. One of those is José Ramos-Horta, leader of the now practically defunct (East) Timor Liberation Front (FRETILIN), who states that "Most wars of national liberation end in negotiations. There is almost no precedent of a guerrilla movement defeating an established government by military means alone. Vietnam and Nicaragua are two of the most recent examples." José Ramos-Horta, *Funu: The Unfinished Saga of East Timor* (Trenton, NJ: The Red Sea Press, 1987), 206.

2. The term *Shining Path* is widely used by everyone in Peru and abroad, but never by the organization's own members. The latter always refer to themselves as members of the Communist party of Peru.

3. It is significant that the anticommunist insurgents today, whether in Angola, Cambodia, Afghanistan, or Nicaragua, see their struggle as directed primarily against foreign (Cuban, Soviet, Vietnamese) occupation. Once foreign troops depart, they fully expect victory.

4. Central Committee, Communist Party of Peru, *Develop the People's War to Serve the World Revolution* (Berkeley, CA: Committee to Support the Revolution in Peru, May 1988), 143.

5. Interview with ERP leader Joaquín Villalobos, in Mario Menendez Rodriguez, *El Salvador: Una Autentica Guerra Civil* (San José: Editorial Universitaria Centroamericana, 1981), 143.

6. "Una experiencia de guerrilla urbana," *Actas Tupamaras*, (Madrid: Editorial Revolución, 1982), 47.

7. Manifesto of the People's Revolutionary Movement, in Foreign Broadcast Information Service, Latin America *(FBIS-LAM)*, 28 December 1982, p. P10.

8. For a thorough critique of liberation theology and its antidemocratic, anticapitalist nature, see Michael Novak, *Will It Liberate? Questions about Liberation Theology* (New York: Paulist Press, 1986).

9. Very few westerners realize that Saloth Sar (a.k.a. Pol Pot) was a student at the Sorbonne; Hafizullah Amin, the communist ruler of Afghanistan killed by the Soviets in December 1979 was a Columbia University graduate; Luis Carrion, a member of the Sandinista nine-member National Directorate, is a graduate of Phillips Exeter Academy and Rensselaer Polytechnic Institute; or that Ethiopian dictator Mengistu Haile Mariam underwent training in the United States. Abimael Guzmán Reynoso, the founder and absolute leader of the Shining Path, wrote his Ph.D. dissertation at the University of Arequipa on Immanuel Kant's concept of time.

10. For an analysis of revolutionary elites in Latin America, see Michael Radu, "Revolution and Revolutionaries," and "Revolutionary Elites," in Michael Radu. editor, *Violence and the Latin American Revolutionaries*, (New Brunswick, NJ: Transaction Books, 1988), 1-14, 108-25.

11. "FPL: The Revolution Is the Best Medium for the Comparison of Strategic and Tactical Approaches (1978)," in Robert S. Leiken and Barry Rubin, eds., *The Central American Crisis Reader* (New York: Summit Books, 1987), 369.

12. PCP (Communist Party of Peru), Central Committee, "Don't Vote—Step Up Guerrilla Warfare to Seize Political Power for the People," *A World to Win*, 3 (1985): 55, 57.

13. Compare Martin Ebon, *The Soviet Propaganda Machine* (New York: McGraw-Hill, 1987), 7-8.

14. Savimbi was long supported by China; he met and was impressed by Mao Zedong; Massoud is known to be a serious student of Maoist and Vietnamese guerrilla tactics, including the role of political warfare; Dhlakhama was a FRELIMO member and presumably indoctrinated in Marxism Leninism, as was his main representative abroad, Artur da Fonseca, who was trained in East Germany for ten years. The Eritreans, Tigreans, and the Khmer Rouge remain, discreetly or not, committed to Marxism-Leninism.

15. It is important to note here that RENAMO, until 1987 perhaps the most militarily successful anticommunist insurgency, has also proved to be the most inept in terms of propaganda, public relations, or sheer presence abroad. On the other side of the coin, the Khmer Rouge and the EPLF, Marxist-Leninist as they are (although in the latter case there seem to be second thoughts), have not managed to shed their image, despite strong incentives and feeble attempts to do so.

16. Carlos Marighella, *Manual of the Urban Guerrilla*, Newly translated with an introduction and a bibliography by Gene Hanrahan (Chapel Hill, N.C.: Documentary Publications, 1985), 84.

17. For an analysis of the important distinctions between terrorism and the use of terror means by bona fide revolutionary organizations, see Michael Radu, "Terror, Terrorism and Insurgency in Latin America," *Orbis* (Spring 1984); also see, Walter Laqueur and Yonah Alexander, eds., *The Terrorism Reader* (New York: Penguin Books, 1987), 298-302.

18. Marighella, *Manual of the Urban Guerrilla,* 81.

19. *Actas Tupamaras*, 16.

20. Menendez, *El Salvador,* 77.

21. Ibid, 138-139.

22. See the text of MRP's manifesto in *La Estrella de Panama*, 22 December 1982.

23. "Isabel", in Menendez, *El Salvador,* 78. Gardner was killed by the FPL; in response, the Salvadoran government broke diplomatic ties with South Africa!

24. For details, see Fred Bridgland, *Jonas Savimbi: A Key to Africa* (New York: Paragon House, 1987), 397-400.

25. For the first-hand testimony of one of the British hostages, see by Glen Dixon as told to Anthony Mockler, *Hostage* (Alberton, South Africa: Galago Press, 1986).

26. Joint Publications Research Services, Latin America *(JPRS-LAM)*, March 1971, 43-44.

27. A single kidnapping brought $60 million to the Montoneros, and FARN was rich enough to be able to "contribute" to the Sandinistas in 1978.

28. Statement in *Barricada*, 5 February 1989, 7.

29. For details on the Cinchoneros, see Michael Radu and Vladimir Tismaneanu, *Revolutionary Movements in Latin America: A Compendium*, (New York: Pergamon Press, forthcoming).

30. See, for instance, "Hondurans Shaken by Others' Wars," *The Christian Science Monitor*, 16 February 1989.

31. L. Francis Bouchey and Alberto M. Piedra, *Guatemala: A Promise in Peril*, (Washington, D.C.: Council for Inter-American Security, 1980), 61-62.

32. Michael Radu, "El Salvador," in Richard Staar, ed. *Yearbook on International Communist Affairs/ 1986*, (Stanford, CA: Hoover Institution Press, 1987), 90.

33. For details on that operation, see Chile: Caos o Democracia. Crónicas de la guerra irregular, no. 2, (Santiago), 1-28.

34. For details on the murder of "Bravo," see Christopher Dickey, *With the Contras. A Reporter in the Wilds of Nicaragua*, (New York: Simon and Schuster, 1985), 64-66; on the killing of Jorge Salazar (no relation to Pablo Emilio), see Shirley Christian, *Nicaragua: Revolution in the Family*, (New York: Random House, 1985), 1871-183; for the background on Somoza's murder in Asuncion, see the claim of former anti-Somoza fighter and Argentine ERP leader, Enrique Gorriarán Merlo; of planning Somoza's murder, see Martin Edwin Andersen, "Dirty Secrets of the 'Dirty War'," *The Nation*, 13 March 1989, 339-346.

35. Text in *FBIS-LAM*, 24 February 1989, 26.

36. Despite the abundance of exposées of Contra "terror" tactics against civilians, none came from an organization or individual who was not paid for, supported by, or sympathetic to the Managua regime. For a thorough debunking of specific incidents, see L. Francis Bouchey, editor, chapters by J. Michael Waller and Steve Baldwin, *The Real Secret War. Sandinista Political Warfare and Its Effects on Congress*, (Washington, D.C.: Council for Inter-American Security, 1987). Even the much maligned *Psychological Operations in Guerrilla Warfare* (New York: Vintage Books, 1985), the so-called CIA manual for the Contras, actually recommends mostly nonviolent propaganda techniques, highly selective and mostly defensive use of weapons against civilians, and certainly no permanent, actual, or extensive use of terror.

37. *FBIS-LAM*, 31 October 1988, 22.

38. *FBIS-LAM*, 24 February 1989, 25.

39. Ibid., 13 February 1989, 26.

40. Ibid., 13 February 1989, 26-27.

41. *Le Monde*, 21 February 1989.

42. Simple arithmetic should demonstrate that the Contras cannot be Somocistas. Indeed, the Somoza National Guard only had some 7000 members before the fall of 1978, and the doubling of that figure at the time only implies the addition of raw recruits and support personnel. Even if the National Guard strength by July 1979 was 14,000 most were untested combatants. Furthermore, the Contras reached a total strength of over 20,000 by 1987, most of whom could not conceivably have been Somocistas in light of the Sandinistas' own claims of killing or jailing some 80 percent of the National Guard. To claim that the Contras are Somocistas simply flies against common sense and is contradicted by everything known about the Nicaraguan resistance. Most of the latter's leaders, and an even larger proportion of rank and file combatants, are either peasants or former Sandinista supporters. Michael Radu, "Los orígenes y la evolución de la insurrección nicarag ense: 1979-1985," *Revista Occidental, Tijuana*, año 4, no. 1 (1987): 67-92.

43. Radio Venceremos, in *FBIS-LAM*, 29 November 1988, 14.

44. See for instance L. Francis Bouchey et al., *The Real Secret War.* Also, a revealing case is that of Michael McClintock, author of a two-volume leftist propaganda book on El Salvador and Guatemala, both subtitled *State Terror and Popular Resistance*. (London: Zed Books, 1985). McClintock is a senior staffer of Amnesty International.

45. Even there, the Permanent Commission on Human Rights (CPDH) and the January 22 Movement of Mothers of Political Prisoners are far more interested in government abuses than of those of the insurgents. Interestingly, they were both patterned after similar groups in El Salvador, Argentina, and Guatemala, all controlled by the far Left.

46. Michael Radu, "El Salvador," in Richard Staar, ed., *Yearbook on International Communist Affairs/1986* (Stanford, CA: Hoover Institution Press), 92.

47. Many such phantom groups are formally led by women alleged to be interested in the fate of their children—often yet another public relations stunt playing on universal human emotions.

48. FMLN control of Honduran refugee camps in Honduras led the Doctors Without Borders, a French relief organization, to pull out in protest in 1988. Guerrilla control is total, and includes penitentiary areas for reeducation. *FBIS-LAM*, 6 December 1988, 26; Mark A. Uhlig, "Fear rules the Salvadoran refugees," *New York Times*, 3 March 1989.

49. Statement by Hector Oqueli Colindres, *FBIS-LAM*, 9 December 1988, 20.

50. The 1987 congressional demand that 5 percent of the $100 million in aid to the Contras be used for a Human Rights Commission may have been the most extreme such instance, but it proves the main argument.

51. Peru is an excellent example in this respect, as both major guerrilla groups have their own legal journalistic mouthpieces: *Cambio* for the Tupac Amarau Revolutionary Movement (MRTA), and *Diario de Marka* for Sendero Luminoso.

52. A typical such call was that of then Archbishop Oscar Romero of El Salvador. See text in "Oscar Romero: The last sermon," in Leiken and Rubin, *The Central American Crisis Reader,* 380.

53. Widely documented are the activities of Marxist-Leninist insurgents on the campuses of such major, and usually public universities as San Carlos in Guatemala, San Marcos in Peru, the National and Central American universities in El Salvador, and the University of the Philippines. It is an almost universal pattern for Marxist-Leninist guerrilla cadres to be recruited from among university students, graduates, and faculty. Teachers' unions throughout Latin America have played a major role in this respect. A particularly insignificant example

is that of Sendero Luminoso's founder and absolute leader, Abimael Guzmán Reynoso, who recruited the entire cadre of his organization at the time when he served as personnel director at the University of San Cristóbal de Huamanga in Ayacucho during the 1970s.

54. For details, see Gabriel Zaid, "Enemy Colleagues," in *Dissent* (Winter 1982).

55. Communiqué of ERP's General Command, in *FBIS-LAM*, 15 March 1983, p. P12.

56. The FMLN's Radio Venceremos even applauded the disaster of the Challenger Space Shuttle on the grounds that two crew members, former Vietnam veterans, were war criminals.

57. For the relevant documents, see Uri Ra'anan et al., *Hydra of Carnage* (Lexington, MA: Lexington Books, 1986), 345-58.

58. Commander "Pedro" in *FBIS-LAM*, 4 June 1980, 9 (annex).

59. See, for instance, Clayton Jones, "A Diplomatic Battle for Legitimacy," *The Christian Science Monitor*, 1 March 1989.

60. UNITA, RENAMO in Mozambique, the Mujahedeen, and the noncommunist Cambodian groups all have representatives abroad, and some have hired public relations experts.

61. Ana Guadalupe Martinez represents ERP on the FDR Political-diplomatic commission, while Mercedes del Carmen Letona runs Radio Venceremos and Ana Sonia Medina Arriola is in charge of internal indoctrination and propaganda.

62. It is significant that when the USIA established a Boston University-led program of training Afghan guerrillas in Peshawar to record combat scenes, all established U.S. media attacked it, and U.S. law prevented that unique footage to reach U.S. audiences.

Selected Bibliography

Alexander G. *A Biographical Directory of 100 Leading Soviet Officials.* Munich: Radio Liberty Research, 1986.

Alinin, S. F., B. G. Antonov, and A. N. Itskov. *The Jonestown Carnage: A CIA Crime.* Moscow: Progress Publishers, 1987.

Barber, Willard F., and C. Neale Ronning. *Internal Security and Military Power: Counterinsurgency and Civic Action in Latin America.* Columbus: Ohio State University Press, 1966.

Barclay, C. N. "Countermeasures against the Urban Guerilla." *Military Review* (January 1972).

Barnett, Frank R., Hugh B. Tovar, and Richard H. Shultz. *Special Operations in US Strategy.* Washington, DC: National Defense University Press, 1984.

Bittman, Ladislav. *The KGB and Soviet Disinformation: An Insider's View.* New York: Pergamon-Brassey, 1985.

Bloomfield, Lincoln P., and Amelia C. Leiss. *Controlling Small Wars.* New York: Knopf, 1969.

Bordzilowski, Jerzy. "Rola Armii Radzieckiej w tworzeniu Ludowego Wojska Poiskiego i w wyzoleniu Polski." *Wojskowy Przeglad Historyczny,* No. 3 (1978).

Bouchey, Francis L., and Alberto M. Piedra. *Guatemala: A Promise in Peril.* Washington DC: Council for Inter-American Security, 1980.

Bridgland, Fred. *Jonas Savimbi: A Key to Africa.* New York: Paragon House, 1987.

Callwell, Sir Charles E. *Small Wars: Their Principles and Practice.* London: Harrison and Sons, 1896.

Cannon, James P. *The Road to Peace.* New York: Pioneer Publishers, 1951.

Chandra, Romesh. *The Present International Situation and the Main Tasks of the Peace Movement; the Role of the World Peace Council Forms and Methods of WPC Activities in the New Conditions.* Sofia: World Peace Council, April 24-27, 1986.

Chernenko, Konstantin. "Topical Questions of the Party's Ideological and Mass Political Work." *Information Bulletin.* (August 1983).

Christian, Shirely. *Nicaragua. Revolution in the Family.* New York: Random House, 1985.

The Communist International, Organ of the Executive Committee of the Communist International. 20 September 1935.

CPSU: Documents and Analysis. 4 February 1988.

Dailey, Brian D., and Patrick J. Parker (eds.). *Soviet Strategic Deception.* Lexington, MA: Lexington Books, 1987.

Davies, Norman. *God's Playground: A History of Poland,* Vol. II. New York: Columbia University Press, 1982.

Deitchman, Seymour J. *Limited War and American Defense Policy.* Cambridge, MA: MIT Press, 1969.

Dickey, Christopher. *With the Contras. A Reporter in the Wilds of Nicaragua.* New York: Simon and Schuster, 1985.

Dolgopolova, Natalya A., and Andrey A. Kokoshin. "Lessons Learned from the Destinies of Great Powers." *Kommunist,* no. 2 (January 1988).

Ebon, Martin. *The Soviet Propaganda Machine*. New York: McGraw-Hill, 1987.

Flashes from the Trade Unions. Prague: World Federation of Trade Unions, 8 July 1988.

Fussell, Paul. *Wartime: Understanding and Behavior in the Second World War.* New York: Oxford University Press, 1989.

Galiullin, Rustem. *The CIA in Asia, Covert Operations Against India and Afghanistan.* Moscow: Progress Publishers, 1988.

Gorbachev, Mikhail. "Address at the UN." Moscow: *Novosti,* 1988.

_______. "For the Sake of Preserving Human Civilization." Moscow: *Novosti,* 1987.

_______. "Restructuring—A Vital Concern of the People." Moscow: *Novosti,* 1987.

_______. "Political Report of the CPSU Central Committee to the 27th Party Congress." Moscow: *Novosti,* 1986.

Gray, Colin. *The Geopolitics of the Nuclear Era: Heartland, Rimlands, and the Technological Revolution.* New York: Crane, Russack and Company, 1979.

Gross, Jan T. *Revolution from Abroad; The Soviet Conquest of Poland's Western Ukraine and Western Belorussia.* Princeton, NJ: Princeton University Press, 1988.

Heilbrunn, Otto. *Warfare in the Enemy's Rear.* New York: Praeger, 1964.

Institute of Social Sciences. *Social Psychology and Propaganda.* Moscow: Progress Publishers, 1987.

*International Affairs (*November 1988): 20-22, 34-37, 40, 49, 57-58, 60.

_______. (October 1988): 3.

Jaubert, Alain. *Making People Disappear: An Amazing Chronicle of Photographic Deception.* Washington, DC: Pergamon-Brassey, 1989.

Jurek, M. and M. Skrzypkowski. *Tarcza Pokoju: Ixlecie Ukladu Warsza Wskiego.* Warsaw: MON, 1975.

Kitson, Frank. *Low Intensity Operations: Subversion, Insurgency, Peacekeeping.* London: Faber, 1971.

Komorowski, T. Bor. *The Secret Army.* New York: Macmillan, 1951.

Korbonski, Stefan. *The Polish Underground State, A Guide to the Underground, 1939-1945.* New York: Columbia University Press, 1978.

Laqueur, Walter. *Guerilla: A Historical and Critical Study.* Boston, MA: Little, Brown, 1976.

Ledeen, Michael, and Herbert Romerstein. *Grenada Documents: An Overview and Selection,* document 28. Washington, DC: U.S. Department of State and Department of Defense, 1984.

Lewis, Bernard. *The Assassins: A Radical Sect in Islam.* London: Weidenfeld & Nicolson, 1967.

Lewy, Guenter. "Deception and Revolutionary War in Vietnam." David Charters and Maurice Tugwell (eds.), *Deception Operations: Studies in the East-West Context.* London, New York: Pergamon-Brassey, forthcoming.

Marighella, Carlos. *Manual of the Urban Guerrilla.* Chapel Hill, NC: Documentary Publications, 1985.

Mason, David S. *Public Opinion and Political Change in Poland, 1980-1982.* Cambridge, MA: Cambridge University Press, 1985.

The Military Balance. London: IISS, 1989.

New Times (Moscow) no. 9 (28 February-6 March 1989).

Nikolayev, Konstantin. "200 Years of the U.S. Constitution—Ideals and Reality." Moscow: *Novosti,* 1987.

Noelle-Neumann, Elizabeth. *Die NATO-Experten und das Publikum: Das Problem der Akzeptanz von Verteidigungspolitik.* Allensbach: Institute for Demoskopie, 1988.

Novak, Michael. *Will It Liberate. Questions about Liberation Theology.* New York: Paulist Press, 1986.

Nowak, Stefan. "Values and Attitudes of the Polish People." *Scientific American* (July 1981).

Nowiecki, Maciej. "W cieniu na wulkanie." *Aneks.* 49 (1988).

Polonsky, A., and B. Drukier. *The Beginnings of Communist Rule in Poland.* London: Routledge and Kegan Paul, 1980.

Radu, Michael. "El Salvador," Richard Staar (ed.), *Yearbook on International Communist Affairs/ 1986.* Stanford, CA: Hoover Institution Press, 1987.

_______. "Revolution and Revolutionaries" and "Revolutionary Elites," in Michael Radu (ed.), *Violence and the Latin American Revolutionaries.* New Brunswick, NJ: Transaction Books, 1988.

_______. "Terror, Terrorism and Insurgency in Latin America." *Orbis* (Spring 1984).

Radu, Michael, and Vladimir Tismaneanu. *Revolutionary Movements in Latin America: A Compendium.* New York: Pergamon Press, forthcoming.

Radvanyi, Janos. *Hungary and the Superpowers: The 1956 Revolution and Realpolitik.* Stanford, CA: Hoover Institution Press, 1972.

Sarkesian, Sam C., and W. L. Scully (eds.). *U.S. Policy and Low Intensity Conflict: Potentials for Military Struggles in the 1980's.* New Brunswick, NJ: Transaction Books, 1981.

Schlacter, David C., and Fred J. Stubbs, "Special Operations Forces: Not Applicable?" *Military Review* (February 1978).

Scott, Harriet Fast, and William F. Scott. *The Armed Forces of the USSR.* Boulder, CO: Westview Press, 1979.

Semiriage, I., A. V. Antosiak, and P. M. Derevianko. *Zarozhdenie Narodnykh Stranuchastnits Varshavskogo Dogovora, 1941-1945.* Moscow: Nauka, 1975.

Shtemenko, S. *Generalnyi Shtab V Gody Voiny.* Moscow: Voenizdat, 1968.

Sobolevf, A.I. *Outline History of the Communist International.* Moscow: Progress Publishers, 1971.

Szabo, Laszlo. "Statement of Laszlo Szabo in Hearings before the CIA Subcommittee of the Armed Services of the House of Representatives, Eighty-ninth Congress." Washington, DC: U.S. Government Printing Office, 1966.

Tugwell, Maurice. "Guilt Transfer," in David C. Rapoport and Yonah Alexander (eds.), *The Morality of Terrorism.* New York: Pergamon Press, 1982.

_______. "Terrorism and Propaganda: Problem and Response," in P. Wilkinson and A. M. Stewart (eds.), *Contemporary Research on Terrorism.* Aberdeen: Aberdeen University Press, 1987.

Valenta, J. *Soviet Intervention in Czechoslovakia 1968: Anatomy of a Decision.* Baltimore, MD: Johns Hopkins University Press, 1979.

Vestnik Ministerstva Inostrannykh Del. 1 December 1988.

Whetten, Lawrence. "The Soviet-Cuban Presence in the Horn of Africa." *RUSI: Journal for Defense Studies* (September 1978).

The World Federation of Trade Unions 1945-1985. Prague: WFTU, October, 1985.

World Marxist Review (September 1988).

World Marxist Review (March 1988).

Yakovlev, Yegor. *Openness Democracy Responsibility.* Moscow: *Novosti,* 1987.

Zagoria, Donald S. "Into the Breach: New Soviet Alliances in the Third World." *Foreign Affairs* (Spring 1979).

Zaid, Gabriel. "Enemy Colleagues." *Dissent* (Winter 1982).

Index

About the Editor and Contributors

Janos Radvanyi, formerly a member of the Hungarian Diplomatic Service whose last post was Chief of the Hungarian Mission in the United States, is presently Director of the Center for International Security and Strategic Studies and the Japan Program of the Center at Mississippi State University. For the past 18 years he has taught East Asian and Contemporary Soviet History at the university. He is an affiliated member of the American and International Associations for the Advancement of Slavic Studies; a member of the International Institute for Strategic Studies, London; and an academic associate of the Atlantic Council of the United States, Washington, D.C. Dr. Radvanyi has lectured widely and is the author and editor of several books and articles, including *Delusion and Reality: Gambits, Hoaxes and Diplomatic One-upmanship in Vietnam; Hungary and the Superpowers;* and *The Pacific in the 1990s: Economic and Strategic Change.*

William E. Odom retired from the U.S. Army in 1988 as a lieutenant general after 34 years of service and assumed his current position as director of National Security Studies for the Hudson Institute. At the time of his retirement, he was director of the National Security Agency and chief of Central Security Service, Ft. George Meade, Maryland, with responsibilities for U.S. signals intelligence and communications security and principal signals intelligence adviser to the secretary of Defense and the director of the Central Intelligence Agency. General Odom has written widely on Soviet political and military affairs, including the book, *The Soviet Volunteers*, and articles in such journals as *World Politics, Foreign Policy, Problems of Communism,* and *The Washington Quarterly*. He is a member of the Council on Foreign Relations and the International Institute for Strategic Studies.

Alfred H. Paddock, Jr., former Director for Psychological Operations in the office of the Secretary of Defense, is highly experienced in psychological and special operations, having served three combat tours in Southeast Asia with the Special Forces and in numerous politico-military and psychological operations command and staff assignments. He commanded the 6th Psychological Operations Battalion and the 4th Psychological Operations Group, and he served on the faculties of the Army Command and General Staff College and the Army War College, where he chaired the Department of National Security. He is the author of the book, *U.S. Army Special Warfare, Its Origins: Psychological and Unconventional Warfare 1941-1952.*

Herbert Romerstein, recently retired director of the Office to Counter Soviet Active Measures at the United States Information Agency (USIA), joined the USIA in 1983 after serving for 18 years as a professional staff member for the U.S. House of Representatives, From 1978 to 1983, he was on the staff of the House Permanent Select Committee on Intelligence. His writings have appeared in the *National Review* and other publications, and he has been a contributor to numerous books including: *The Red Orchestra, Mesmerized by the Bear,* and *Grenada Documents: An Overview and Selection.* His book, *The KGB Against the Main Enemy—The Story of Soviet Espionage Against the United States*, was published in 1989.

Maurice A. J. Tugwell, a retired British Army Brigadier, established the Mackenzie Institute for the Study of Terrorism, Revolution and Propaganda in 1986. Dr. Tugwell entered the British Army in 1943 as a private and served continuously until his retirement as a Brigadier in 1978. He participated in the Rhine Crossing in 1945 and has served in India, Palestine, Malaya, Cyprus, Aden, Iran, and Northern Ireland. He was the founding director of the Centre for Conflict Studies at the University of New Brunswick, Canada, and his books include: *Peace With Freedom, Arnhem, A Case Study, Airborne to Battle, Armies in Low Intensity Conflict* (joint editor and contributing author), and *Deception in the East-West Context*. His work has also appeared in numerous journals.

Josef Joffe is the foreign editor and columnist of West Germany's *Suddeutsche Zeitung* and contributing editor of *U.S. News and World Report.* Between 1982 and 1984, he taught international relations at the Johns Hopkins School of Advanced International Studies. He was senior editor of the German weekly, *Die Zeit*, from 1976 until 1982. His most recent book is *The Limited Partnership: Europe, the United States, and the Burdens of Alliance.* His writings on international politics and strategy have appeared in *Foreign Affairs, Foreign Policy,* and the

Washington Quarterly. He has also written extensively for the international press, including *The Washington Post, Wall Street Journal, Los Angeles Times, Jerusalem Post, International Herald Tribune,* and *New Republic*.

Teresa Rakowska-Harmstone was a professor of Political Science at Carleton University, Canada, and served as chair of the department. Previously, Dr. Rakowska-Harmstone has been a visiting professor at McGill and George Washington Universities and has worked as a research fellow and/or visiting scholar at the Institute of Sino-Soviet Studies, George Washington University; the Institute of International Politics and Economics, Belgrade; the Hoover Institution, Stanford University; the Russian Research Center, Harvard University; and the Stiftung Wissenchaft und Politik, Ebenhausen, West Germany. She is presently a Secretary of the Navy Fellow at the U.S. Naval Academy. Among other publication credits, she is editor and coauthor of *Communism in Eastern Europe*.

Pierre Lacoste escaped from occupied France in 1943 and traveled to North Africa via Spain to join the French naval forces. After his admission to the Naval Academy in 1944, he took part in war operations in the Mediterranean and later in Indochina. Prior to his position as president of the Fondation pour les Etudes de Defense National (FEDN), Admiral Lacoste served as commander of the Naval War College, military adviser to the prime minister, commander of the Mediterranean Fleet, and head of the Foreign Intelligence Service (DGSE).

Michael Radu, a resident scholar at the Foreign Policy Research Institute (FPRI) in Philadelphia, has also been since 1985 a contributing editor to ORBIS. He has held teaching appointments as a visiting lecturer at the University of Pennsylvania, Temple University, and the University of the Witwatersrand in Johannesburg, South Africa. His books include *Violence and Latin American Revolutionaries; From Power to Irrelevance: A History of the Guatemalan Left; The New Insurgents, Anti-communist Guerrillas in the Third World;* and *Southern Discomfort: Soviet Activities in Sub Saharan Africa.*